RISK MANAGEMENT HANDBOOK

RISK MANAGEMENT HANDBOOK

FAA-H-8083-2

Federal Aviaton Administration

Skyhorse Publishing

Skyhorse Publishing books may be purchased in bulk at special discounts for sales promotion, corporate gifts, fund-raising, or educational purposes. Special editions can also be created to specifications. For details, contact the Special Sales Department, Skyhorse Publishing, 307 West 36th Street, 11th Floor, New York, NY 10018 or info@ skyhorsepublishing.com.

Skyhorse® and Skyhorse Publishing® are registered trademarks of Skyhorse Publishing, Inc.®, a Delaware corporation.

Visit our website at www.skyhorsepublishing.com.

10 9 8 7 6 5 4 3 2 1

Library of Congress Cataloging-in-Publication Data is available on file.

ISBN: 978-1-61608-697-8

Printed in China

Preface

This handbook is a tool designed to help recognize and manage risk. It provides a higher level of training to the pilot in command (PIC) who wishes to aspire to a greater understanding of the aviation environment and become a better pilot. This handbook is for pilots of all aircraft from Weight-Shift Control (WSC) to a Piper Cub, a Twin Beechcraft, or a Boeing 747. A pilot's continued interest in building skills is paramount for safe flight and can assist in rising above the challenges which face pilots of all backgrounds.

Some basic tools are provided in this handbook for developing a competent evaluation of one's surroundings that allows for assessing risk and thereby managing it in a positive manner. Risk management is examined by reviewing the components that affect risk thereby allowing the pilot to be better prepared to mitigate risk.

The pilot's work requirements vary depending on the mode of flight. As for a driver transitioning from an interstate onto the city streets of New York, the tasks increase significantly during the landing phase, creating greater risk to the pilot and warranting actions that require greater precision and attention. This handbook attempts to bring forward methods a pilot can use in managing the workloads, making the environment safer for the pilot and the passengers. *[Figure I-1]*

This handbook may be purchased from the Superintendent of Documents, United States Government Printing Office (GPO), Washington, DC 20402-9325, or from the GPO website at http://bookstore.gpo.gov.

This handbook is also available for download, in PDF format, from the Regulatory Support Division (AFS-600) website at http://www.faa.gov.

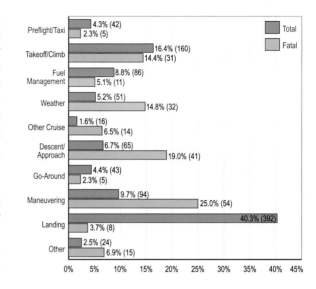

Figure I-1. *The percentage of aviation accidents by phase of flight.*

Occasionally, the word "must" or similar language is used where the desired action is deemed critical. The use of such language is not intended to add to, interpret, or relieve a duty imposed by Title 14 of the Code of Federal Regulations (14 CFR).

Comments regarding this publication should be sent, in email form, to the following address:

AFS630comments@faa.gov

Introduction

According to National Transportation Board (NTSB) statistics, in the last 20 years, approximately 85 percent of aviation accidents have been caused by "pilot error." Many of these accidents are the result of the tendency to focus flight training on the physical aspects of flying the aircraft by teaching the student pilot enough aeronautical knowledge and skill to pass the written and practical tests. Risk management is ignored, with sometimes fatal results. The certificated flight instructor (CFI) who integrates risk management into flight training teaches aspiring pilots how to be more aware of potential risks in flying, how to clearly identify those risks, and how to manage them successfully.

"A key element of risk decision-making is determining if the risk is justified."

The risks involved with flying are quite different from those experienced in daily activities. Managing these risks requires a conscious effort and established standards (or a maximum risk threshold). Pilots who practice effective risk management have predetermined personal standards and have formed habit patterns and checklists to incorporate them.

If the procedures and techniques described in this handbook are taught and employed, pilots will have tools to determine the risks of a flight and manage them successfully. The goal is to reduce the general aviation accident rate involving poor risk management. Pilots who make a habit of using risk management tools will find their flights considerably more enjoyable and less stressful for themselves and their passengers. In addition, some aircraft insurance companies reduce insurance rates after a pilot completes a formal risk management course.

This Risk Management Handbook makes available recommended tools for determining and assessing risk in order to make the safest possible flight with the least amount of risk. The appendices at the end of this handbook contain checklists and scenarios to aid in risk management consideration, flight planning, and training.

Acknowledgments

The Risk Management Handbook was produced by the Federal Aviation Administration (FAA) with the assistance of Safety Research Corporation of America. The FAA wishes to acknowledge the following contributors:

Dr. Pat Veillette for information used on human behaviors (chapter 2)

Cessna Aircraft Company and Garmin Ltd. for images provided and used throughout the Handbook

Additional appreciation is extended to the Aircraft Owners and Pilots Association (AOPA), the AOPA Air Safety Foundation, and the National Business Aviation Association (NBAA) for their technical support and input.

Table of Contents

Preface..iii

Introduction...v

Acknowledgments...vii

Table of Contents ix

Chapter 1
Defining Elements of Risk Management 1-1
Introduction...1-1
   Hazard 1-2
   Risk...1-5
Managing Risks 1-5
Chapter Summary 1-8

Chapter 2
Human Behavior 2-1
Introduction...2-1
Chapter Summary 2-5

Chapter 3
Identifying Hazards and Mitigating Risk............3-1
Introduction...3-1
P = Pilot in command 3-3
   The Pilot's Health.....................................3-3
   Stress Management....................................3-4
A = Aircraft 3-4
V = Environment..3-5
   Weather 3-5
   Terrain 3-5
   Airport 3-6
   Airspace..3-6
   Nighttime..3-6

   Visual Illusions..3-7
E = External Pressures 3-9
Chapter Summary 3-9

Chapter 4
Assessing Risk...4-1
Introduction...4-1
Quantifying Risk Using a Risk Matrix 4-2
   Likelihood of an Event............................4-2
   Severity of an Event 4-2
   Mitigating Risk...4-4
Chapter Summary 4-4

Chapter 5
Aeronautical Decision-Making:
A Basic Staple..5-1
Introduction...5-1
History of ADM..5-2
Analytical Decision-Making...........................5-3
Automatic Decision-Making...........................5-4
Operational Pitfalls 5-4
   Scud Running 5-6
   Get-There-Itis...5-6
   Continuing VFR into IMC 5-7
   Loss of Situational Awareness 5-8
   Flying Outside the Envelope 5-9
3P Model 5-10
   Rate of Turn..5-10
   Radius of Turn..5-11
   Perceive 5-11
   Process..5-13
   Perform...5-13
Chapter Summary 5-13

Chapter 6
Single-Pilot
Resource Management **6-1**
Introduction...6-1
Recognition of Hazards...6-2
Use of Resources...6-3
    Internal Resources...6-6
    External Resources..6-8
SRM and the 5P Check 6-8
    Plan 6-11
    Plane 6-11
    Pilot 6-12
    Passengers 6-12
    Programming 6-13
Chapter Summary 6-14

Chapter 7
Automation...**7-1**
Introduction...7-1
Cockpit Automation Study 7-3
Realities of Automation 7-4
Enhanced Situational Awareness............................7-6
Autopilot Systems...7-8
    Familiarity 7-8
    Respect for Onboard Systems 7-8
    Reinforcement of Onboard Suites.....................7-8
    Getting Beyond Rote Workmanship 7-8
    Understand the Platform 7-8
Flight Management Skills..7-9
    Automation Management 7-9
    Information Management...................................7-9
    Risk Management..7-10
Chapter Summary 7-10

Chapter 8
Risk Management Training................................**8-1**
Introduction...8-1
System Safety Flight Training 8-2
Setting Personal Minimums....................................8-3
    Step 1—Review Weather Minimums.................8-3
    Step 2—Assess Experience and Comfort Level.........8-3
    Step 3—Consider Other Conditions..................8-5
    Step 4—Assemble and Evaluate 8-5
    Step 5—Adjust for Specific Conditions 8-6
    Step 6—Stick to the Plan! 8-6
Chapter Summary 8-7

Appendix A
Personal Assessment and
Minimums...**A-1**

Appendix B
Sample Risk Management Scenarios..............**B-1**

Appendix C
CFIT Checklist...**C-1**

Glossary **G-1**

Index **I-1**

Chapter 1

Defining Elements of Risk Management

Introduction

Risk management, a formalized way of dealing with hazards, is the logical process of weighing the potential costs of risks against the possible benefits of allowing those risks to stand uncontrolled. In order to better understand risk management, the terms "hazard" and "risk" need to be understood.

Attitude
Training
Education
Predisposition
Background

PILOT 1

Attitude
Training
Education
Predisposition
Background

PILOT 2

Education Attitude
Back

P

Types of Risk

Total Risk	The sum of identified and unidentified risks.
Identified Risk	Risk which has been determined through various analysis techniques. The first task of system safety is to identify, within practical limitations, all possible risks.
Unidentified Risk	Risk not yet identified. Some unidentified risks are subsequently identified when a mishap occurs. Some risk is never known.
Unacceptable Risk	Risk which cannot be tolerated by the managing activity. It is a subset of identified risk that must be or controlled.
Acceptable Risk	Acceptable risk is the part of identified risk that is allowed to further engineering action. Making this difficult yet necessary the managing a made with full user who is e
Residual Risk	Residual r system s employe same risk is

Risk Assessment Matrix

Likelihood	Severity			
	Catastrophic	Critical	Marginal	Negligible
Probable	High	High	Serious	
Occasional	High	Serious		
Remote	Serious	Medium		Low
Improbable				

Hazard

Defining Hazard

By definition, a hazard is a present condition, event, object, or circumstance that could lead to or contribute to an unplanned or undesired event such as an accident. It is a source of danger. Four common aviation hazards are:

1. A nick in the propeller blade
2. Improper refueling of an aircraft
3. Pilot fatigue
4. Use of unapproved hardware on aircraft

Recognizing the Hazard

Recognizing hazards is critical to beginning the risk management process. Sometimes, one should look past the immediate condition and project the progression of the condition. This ability to project the condition into the future comes from experience, training, and observation.

1. A nick in the propeller blade is a hazard because it can lead to a fatigue crack, resulting in the loss of the propeller outboard of that point. With enough loss, the vibration could be great enough to break the engine mounts and allow the engine to separate from the aircraft.

2. Improper refueling of an aircraft is a hazard because improperly bonding and/or grounding the aircraft creates static electricity that can spark a fire in the refueling vapors. Improper refueling could also mean fueling a gasoline fuel system with turbine fuel. Both of these examples show how a simple process can become expensive at best and deadly at worst.

3. Pilot fatigue is a hazard because the pilot may not realize he or she is too tired to fly until serious errors are made. Humans are very poor monitors of their own mental condition and level of fatigue. Fatigue can be as debilitating as drug usage, according to some studies.

4. Use of unapproved hardware on aircraft poses problems because aviation hardware is tested prior to its use on an aircraft for such general properties as hardness, brittleness, malleability, ductility, elasticity, toughness, density, fusibility, conductivity, and contraction and expansion.

If pilots do not recognize a hazard and choose to continue, the risk involved is not managed. However, no two pilots see hazards in exactly the same way, making prediction and standardization of hazards a challenge. So the question remains, how do pilots recognize hazards? The ability to recognize a hazard is predicated upon personality, education, and experience.

Personality

Personality can play a large part in the manner in which hazards are gauged. People who might be reckless in nature take this on board the flight deck. For instance, in an article in the August 25, 2006, issue of Commercial and Business Aviation entitled Accident Prone Pilots, Patrick R. Veillette, Ph.D., notes that research shows one of the primary characteristics exhibited by accident-prone pilots was their disdain toward rules. Similarly, other research by Susan Baker, Ph.D., and her team of statisticians at the Johns Hopkins School of Public Health, found a very high correlation between pilots with accidents on their flying records and safety violations on their driving records. The article brings forth the question of how likely is it that someone who drives with a disregard of the driving rules and regulations will then climb into an aircraft and become a role model pilot. The article goes on to hypothesize that, for professional pilots, the financial and career consequences of deviating from standard procedures can be disastrous but can serve as strong motivators for natural-born thrill seekers.

Improving the safety records of the thrill seeking type pilots may be achieved by better educating them about the reasons behind the regulations and the laws of physics, which cannot be broken. The FAA rules and regulations were developed to prevent accidents from occurring. Many rules and regulations have come from studying accidents; the respective reports are also used for training and accident prevention purposes.

Education

The adage that one cannot teach an old dog new tricks is simply false. In the mid-1970s, airlines started to employ Crew Resource Management (CRM) in the workplace (flight deck). The program helped crews recognize hazards and provided tools for them to eliminate the hazard or minimize its impact. Today, this same type of thinking has been integrated into Single-Pilot Resource Management (SRM) programs (see chapter 6).

Regulations

Regulations provide restrictions to actions and are written to produce outcomes that might not otherwise occur if the regulation were not written. They are written to reduce hazards by establishing a threshold for the hazard. An example might be something as simple as basic visual flight rules (VFR) weather minimums as presented in Title 14 of the Code of Federal Regulation (14 CFR) part 91, section 91.155, which lists cloud clearance in Class E airspace as 1,000 feet below, 500 feet above, and 2,000 feet horizontally with flight visibility as three statute miles. This regulation provides both an operational boundary and one that a pilot can use in helping to recognize a hazard. For instance, a VFR-only rated pilot faced with weather that is far below that of Class E airspace would recognize that weather as hazardous, if for no other reason than because it falls below regulatory requirements.

Experience

Experience is the knowledge acquired over time and increases with time as it relates to association with aviation and an accumulation of experiences. Therefore, can inexperience be construed as a hazard? Inexperience is a hazard if an activity demands experience of a high skill set and the inexperienced pilot attempts that activity. An example of this would be a wealthy pilot who can afford to buy an advanced avionics aircraft, but lacks the experience needed to operate it safely. On the other hand a pilot's experience can provide a false sense of security, leading the pilot to ignore or fail to recognize a potential hazard.

Experience sometimes influences the way a pilot looks at an aviation hazard and how he or she explores its level of risk. Revisiting the four original examples:

1. **A nick in the propeller blade.** The pilot with limited experience in the field of aircraft maintenance may not realize the significance of the nick. Therefore, he or she may not recognize it as a hazard. For the more experienced pilot, the nick represents the potential of a serious risk. This pilot realizes the nick can create or be the origin of a crack. What happens if the crack propagates, causing the loss of the outboard section? The ensuing vibration and possible loss of the engine would be followed by an extreme out-of-balance condition resulting in the loss of flight control and a crash.

2. **Improper refueling of an aircraft.** Although pilots and servicing personnel should be well versed on the grounding and/or bonding precautions as well as the requirements for safe fueling, it is possible the inexperienced pilot may be influenced by haste and fail to take proper precautions. The more experienced pilot is aware of how easily static electricity can be generated and how the effects of fueling a gasoline fuel system with turbine fuel can create hazards at the refueling point.

3. **Pilot fatigue.** Since indications of fatigue are subtle and hard to recognize, it often goes unidentified by a pilot. The more experienced pilot may actually ignore signals of fatigue because he or she believes flight experience will compensate for the hazard. For example, a businessman/pilot plans to fly to a meeting and sets an 8 a.m. departure for himself. Preparations for the meeting keep him up until 2 a.m. the night before the flight. With only several hours of sleep, he arrives at the airport ready to fly because he fails to recognize his lack of sleep as a hazard. The fatigued pilot is an impaired pilot, and flying requires unimpaired judgment. To offset the risk of fatigue, every pilot should get plenty of rest and minimize stress before a flight. If problems prevent a good night's sleep, rethink the flight, and postpone it accordingly.

4. **Use of unapproved hardware on aircraft.** Manufacturers specify the type of hardware to use on an aircraft, including components. Using anything other than that which is specified or authorized by parts manufacturing authorization (PMA) is a hazard. There are several questions that a pilot should consider that further explain why unapproved hardware is a hazard. Will it corrode when in contact with materials in the airframe structure? Will it break because it is brittle? Is it manufactured under loose controls such that some bolts may not meet the specification? What is the quality control process at the manufacturing plant? Will the hardware deform excessively when torqued to the proper specification? Will it stay tight and fixed in place with the specified torque applied? Is it loose enough to allow too much movement in the structure? Are the dollars saved really worth the possible costs and liability? As soon as a person departs from the authorized design and parts list, then that person becomes an engineer and test pilot, because the structure is no longer what was considered to be safe and approved. Inexperienced as well as experienced pilots can fall victim to using an unapproved part, creating a flight hazard that can lead to an accident. Aircraft manufacturers use hardware that meets multiple specifications that include shear strength, tensile strength, temperature range, working load, etc.

Tools for Hazard Awareness

There are some basic tools for helping recognize hazards.

Advisory Circulars (AC)

Advisory circulars (ACs) provide nonregulatory information for helping comply with 14 CFR. They amplify the intent of the regulation. For instance, AC 90-48, Pilot's Role in Collision Avoidance, provides information about the amount of time it takes to see, react, and avoid an oncoming aircraft.

For instance, if two aircraft are flying toward each other at 120 knots, that is a combined speed of 240 knots. The distance that the two aircraft are closing at each other is about 400 feet per second (403.2 fps). If the aircraft are one mile apart, it only takes 13 seconds (5,280 ÷ 400) for them to impact. According to AC 90-48, it takes a total of 12.5 seconds for the aircraft to react to a pilot's input after the pilot sees the other aircraft. [*Figure 1-1*]

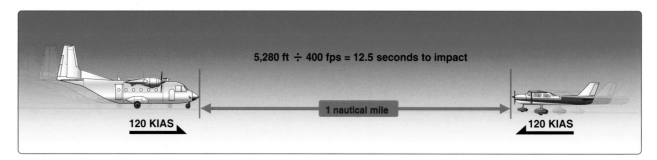

Figure 1-1. *Head-on approach impact time.*

Understanding the Dangers of Converging Aircraft

If a pilot sees an aircraft approaching at an angle and the aircraft's relationship to the pilot does not change, the aircraft will eventually impact. If an aircraft is spotted at 45° off the nose and that relationship remains constant, it will remain constant right up to the time of impact (45°). Therefore, if a pilot sees an aircraft on a converging course and the aircraft remains in the same position, change course, speed, altitude or all of these to avoid a midair collision.

Understanding Rate of Climb

In 2006, a 14 CFR part 135 operator for the United States military flying Casa 212s had an accident that would have been avoided with a basic understanding of rate of climb. The aircraft (flying in Afghanistan) was attempting to climb over the top ridge of a box canyon. The aircraft was climbing at 1,000 feet per minute (fpm) and about 1 mile from the canyon end. Unfortunately, the elevation change was also about 1,000 feet, making a safe ascent impossible. The aircraft hit the canyon wall about ½ way up the wall. How is this determined? The aircraft speed in knots multiplied by 1.68 equals the aircraft speed in feet per second (fps). For instance, in this case if the aircraft were traveling at about 150 knots, the speed per second is about 250 fps (150 X 1.68). If the

aircraft is a nautical mile (NM) (6,076.1 feet) from the canyon end, divide the one NM by the aircraft speed. In this case, 6,000 feet divided by 250 is about 24 seconds. [*Figure 1-2*]

Understanding the Glide Distance

In another accident, the instructor of a Piper Apache feathered the left engine while the rated student pilot was executing an approach for landing in VFR conditions. Unfortunately, the student then feathered the right engine. Faced with a small tree line (containing scrub and small trees less than 10 feet in height) to his front, the instructor attempted to turn toward the runway. As most pilots know, executing a turn results in either decreased speed or increased descent rate, or requires more power to prevent the former. Starting from about 400 feet without power is not a viable position, and the sink rate on the aircraft is easily between 15 and 20 fps vertically. Once the instructor initiated the turn toward the runway, the sink rate was increased by the execution of the turn. [*Figure 1-3*] Adding to the complexity of the situation, the instructor attempted to unfeather the engines, which increased the drag, in turn increasing the rate of descent as the propellers started to turn. The aircraft stalled, leading to an uncontrolled impact. Had the instructor continued straight

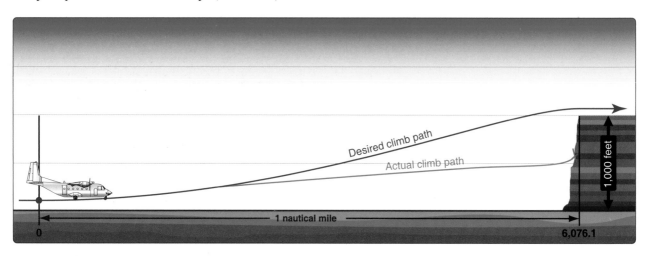

Figure 1-2. *The figure above is a scale drawing of an aircraft climbing at 1,000 fpm, located 1 NM from the end of the canyon and starting from the canyon floor 1,000 feet below the rim. The time to cover 6,000 feet is 24 seconds. With the aircraft climbing at 1,000 fps, in approximately ½ minute, the aircraft will climb only 500 feet and will not clear the rim.*

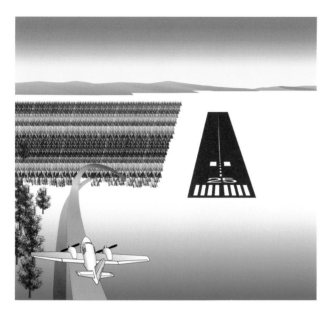

Figure 1-3. *In attempting to turn toward the runway, the instructor pilot landed short in an uncontrolled manner, destroying the aircraft and injuring both pilots.*

ahead, the aircraft would have at least been under control at the time of the impact.

There are several advantages to landing under control:

- The pilot can continue flying to miss the trees and land right side up to enhance escape from the aircraft after landing.

- If the aircraft lands right side up instead of nose down, or even upside down, there is more structure to absorb the impact stresses below the cockpit than there is above the cockpit in most aircraft.

- Less impact stress on the occupants means fewer injuries and a better chance of escape before fires begin.

Risk
Defining Risk
Risk is the future impact of a hazard that is not controlled or eliminated. It can be viewed as future uncertainty created by the hazard. If it involves skill sets, the same situation may yield different risk.

1. If the nick is not properly evaluated, the potential for propeller failure is unknown.

2. If the aircraft is not properly bonded and grounded, there is a build-up of static electricity that can and will seek the path of least resistance to ground. If the static discharge ignites the fuel vapor, an explosion may be imminent.

3. A fatigued pilot is not able to perform at a level commensurate with the mission requirements.

4. The owner of a homebuilt aircraft decides to use bolts from a local hardware store that cost less than the recommended hardware, but look the same and appear to be a perfect match, to attach and secure the aircraft wings. The potential for the wings to detach during flight is unknown.

In scenario 3, what level of risk does the fatigued pilot present? Is the risk equal in all scenarios and conditions? Probably not. For example, look at three different conditions in which the pilot could be flying:

1. Day visual meteorological conditions (VMC) flying visual flight rules (VFR)

2. Night VMC flying VFR

3. Night instrument meteorological conditions (IMC) flying instrument flight rules (IFR)

In these weather conditions, not only the mental acuity of the pilot but also the environment he or she operates within affects the risk level. For the relatively new pilot versus a highly experienced pilot, flying in weather, night experience, and familiarity with the area are assessed differently to determine potential risk. For example, the experienced pilot who typically flies at night may appear to be a low risk, but other factors such as fatigue could alter the risk assessment.

In scenario 4, what level of risk does the pilot who used the bolts from the local hardware center pose? The bolts look and feel the same as the recommended hardware, so why spend the extra money? What risk has this homebuilder created? The bolts purchased at the hardware center were simple low-strength material bolts while the wing bolts specified by the manufacturer were close-tolerance bolts that were corrosion resistant. The bolts the homebuilder employed to attach the wings would probably fail under the stress of takeoff.

Managing Risks

Risk is the degree of uncertainty. An examination of risk management yields many definitions, but it is a practical approach to managing uncertainty. [*Figure 1-4*] Risk assessment is a quantitative value assigned to a task, action, or event. [*Figure 1-5*] When armed with the predicted assessment of an activity, pilots are able to manage and reduce (mitigate) their risk. Take the use of improper hardware on a homebuilt aircraft for construction. Although one can easily see both the hazard is high and the severity is extreme, it does take the person who is using those bolts to recognize the risk. Otherwise, as is in many cases, the chart in *Figure 1-5* is used after the fact. Managing risk takes discipline in separating oneself from the activity at hand in order to view the situation as an unbiased evaluator versus

Types of Risk	
Total Risk	The sum of identified and unidentified risks.
Identified Risk	Risk that has been determined through various analysis techniques. The first task of system safety is to identify, within practical limitations, all possible risks.
Unidentified Risk	Risk not yet identified. Some unidentified risks are subsequently identified when a mishap occurs. Some risk is never known.
Unacceptable Risk	Risk that cannot be tolerated by the managing activity. It is a subset of identified risk that must be eliminated or controlled.
Acceptable Risk	Acceptable risk is the part of identified risk that is allowed to persist without further engineering or management action. Making this decision is a difficult yet necessary responsibility of the managing activity. This decision is made with full knowledge that it is the user who is exposed to this risk.
Residual Risk	Residual risk is the risk remaining after system safety efforts have been fully employed. It is not necessarily the same as acceptable risk. Residual risk is the sum of acceptable risk and unidentified risk. This is the total risk passed on to the user.

Figure 1-4. *Types of risk.*

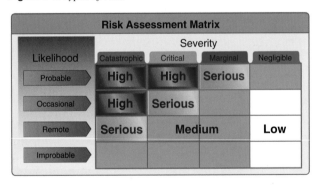

Figure 1-5. *Using a risk assessment matrix helps the pilot differentiate between low-risk and high-risk flights.*

an eager participant with a stake in the flight's execution. Another simple step is to ask three questions—is it safe, is it legal, and does it make sense? Although not a formal methodology of risk assessment, it prompts a pilot to look at the simple realities of what he or she is about to do.

Therefore, risk management is the method used to control, eliminate, or reduce the hazard within parameters of acceptability. Risk management is unique to each and every individual, since there are no two people exactly alike in skills, knowledge, training, and abilities. An acceptable level of risk to one pilot may not necessarily be the same to another pilot. Unfortunately, in many cases the pilot perceives that his or her level of risk acceptability is actually greater than their capability thereby taking on risk that is dangerous.

It is a decision-making process designed to systematically identify hazards, assess the degree of risk, and determine the best course of action. Once risks are identified, they must be assessed. The risk assessment determines the degree of risk (negligible, low, medium, or high) and whether the degree of risk is worth the outcome of the planned activity. If the degree of risk is "acceptable," the planned activity may then be undertaken. Once the planned activity is started, consideration must then be given whether to continue. Pilots must have viable alternatives available in the event the original flight cannot be accomplished as planned.

Thus, hazard and risk are the two defining elements of risk management. A hazard can be a real or perceived condition, event, or circumstance that a pilot encounters.

Consider the example of a flight involving a Beechcraft King Air. The pilot was attempting to land in a northern Michigan airport. The forecasted ceilings were at 500 feet with ½ mile visibility. He deliberately flew below the approach minimums, ducked under the clouds, and struck the ground killing all on board. A prudent pilot would assess the risk in this case as high and beyond not only the capabilities of the aircraft and the pilot but beyond the regulatory limitations established for flight. The pilot failed to take into account the hazards associated with operating an aircraft in low ceiling and low visibility conditions.

A review of the accident provides a closer look at why the accident happened. If the King Air were traveling at 140 knots or 14,177 feet per minute, it would cover ½ statute mile (sm) visibility (2,640 feet) in about 11 seconds. As determined in *Figure 1-1*, the pilot has 12.5 seconds to impact. This example states that the King Air is traveling ½ statute mile every 11 seconds, so if the pilot only had ½ sm visibility, the aircraft will impact before the pilot can react. These factors make flight in low ceiling and low visibility conditions extremely hazardous. Chapter 4, Aerodynamics of Flight, of the Pilot's Handbook of Aeronautical Knowledge presents a discussion of space required to maneuver an aircraft at various airspeed.

So, why would a pilot faced with such hazards place those hazards at such a low level of risk? To understand this, it is important to examine the pilot's past performance. The pilot had successfully flown into this airport under similar conditions as these despite the apparent risk. This time, however, the conditions were forecast with surface fog. Additionally, the pilot and his passenger were in a hurry. They were both late for their respective appointments. Perhaps being in a hurry, the pilot failed to factor in the difference between the forecasted weather and weather he negotiated before. Can it be said that the pilot was in a hurry definitively? Two years before this accident, the pilot landed a different aircraft gear up. At that incident, he simply told the fixed-base operator (FBO) at the airport to take care of the aircraft because the pilot needed to go to a meeting. He also had an enforcement action for flying low over a populated area.

It is apparent that this pilot knew the difference between right and wrong. He elected to ignore the magnitude of the hazard, the final illustration of a behavioral problem that ultimately caused this accident. Certainly one would say that he was impetuous and had what is called "get there itis." While ducking under clouds to get into the Michigan airport, the pilot struck terrain killing everyone onboard. His erroneous behavior resulted from inadequate or incorrect perceptions of the risk, and his skills, knowledge, and judgment were not sufficient to manage the risk or safely complete the tasks in that aircraft. [*Figure 1-6*]

The hazards a pilot faces and those that are created through adverse attitude predispose his or her actions. Predisposition is formed from the pilot's foundation of beliefs and, therefore, affects all decisions he or she makes. These are called "hazardous attitudes" and are explained in the Pilot's Handbook of Aeronautical Knowledge, Chapter 17, Aeronautical Decision-Making.

A key point must be understood about risk. Once the situation builds in complexity, it exceeds the pilot's capability and requires luck to succeed and prevail. [*Figure 1-7*]

Unfortunately, when a pilot survives a situation above his or her normal capability, perception of the risk involved and of the ability to cope with that level of risk become skewed. The pilot is encouraged to use the same response to the same perceived level of risk, viewing any success as due to skill, not luck. The failure to accurately perceive the risk involved and the level of skill, knowledge, and abilities required to mitigate that risk may influence the pilot to accept that level of risk or higher levels.

Many in the aviation community would ask why the pilot did not see this action as a dangerous maneuver. The aviation community needs to ask questions and develop answers to these questions: "What do we need to do during the training and education of pilots to enable them to perceive these hazards as risks and mitigate the risk factors?" "Why was this

Figure 1-6. *Each pilot may have a different threshold where skill is considered, however; in this case no amount of skill raises this line to a higher level.*

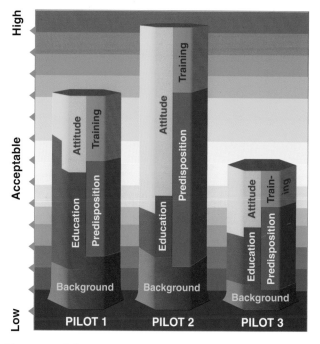

Figure 1-7. *Pilots accept their own individual level of risk even though they may have received similar training. Risk, which must be managed individually, becomes a problem when a situation builds and its complexity exceeds the pilot's capability (background + education + predisposition + attitude + training). The key to managing risk is the pilot's understanding of his or her threshold and perceptions of the risk.*

pilot not trained to ask for an approach clearance and safely fly an approach or turned around and divert to an airport with better weather?" Most observers view this approach as not only dangerous but also lacking common sense. To further understand this action, a closer look at human behavior is provided in Chapter 2, Studies of Human Behavior.

Chapter Summary

The concepts of hazard and risk are the core elements of risk management. Types of risk and the experience of the pilot determine that individual's acceptable level of risk.

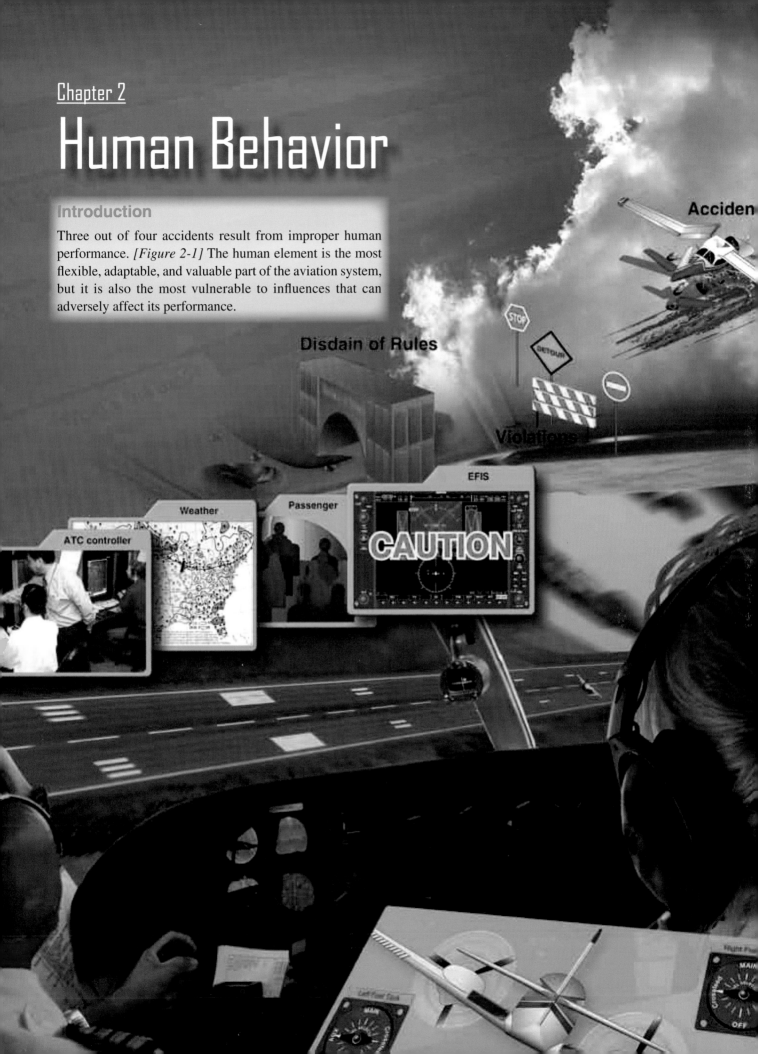

Human Behavior

Introduction

Three out of four accidents result from improper human performance. *[Figure 2-1]* The human element is the most flexible, adaptable, and valuable part of the aviation system, but it is also the most vulnerable to influences that can adversely affect its performance.

Acciden

Disdain of Rules

STOP

DETOUR

Violations

EFIS

Weather

Passenger

ATC controller

CAUTION

Figure 2-1. *Three out of four accidents result from human error.*

The study of human behavior is an attempt to explain how and why humans function the way they do. A complex topic, human behavior is a product both of innate human nature and of individual experience and environment. Definitions of human behavior abound, depending on the field of study. In the scientific world, human behavior is seen as the product of factors that cause people to act in predictable ways.

The Federal Aviation Administration (FAA) utilizes studies of human behavior in an attempt to reduce human error in aviation. Historically, the term "pilot error" has been used to describe an accident in which an action or decision made by the pilot was the cause or a contributing factor that led to the accident. This definition also includes the pilot's failure to make a correct decision or take proper action. From a broader perspective, the phrase "human factors related" more aptly describes these accidents. A single decision or event does not lead to an accident, but a series of events; the resultant decisions together form a chain of events leading to an outcome. Many of these events involve the interaction of flight crews. In fact, airlines have long adopted programs for crew resource management (CRM) and line oriented flight training (LOFT) which has had a positive impact upon both safety and profit. These same processes can be applied (to an extent) to general aviation.

Human error may indicate where in the system a breakdown occurs, but it provides no guidance as to why it occurs. The effort of uncovering why pilots make mistakes is multidisciplinary in nature. In aviation—and with pilots in particular—some of the human factors to consider when examining the human role are decision-making, design of displays and controls, flight deck layout, communications, software, maps and charts, operating manuals, checklists and system procedures. Any one of the above could be or become a stressor that triggers a breakdown in the human performance that results in a critical human error.

Since poor decision-making by pilots (human error) has been identified as a major factor in many aviation accidents, human behavior research tries to determine an individual's predisposition to taking risks and the level of an individual's involvement in accidents. Drawing upon decades of research, countless scientists have tried to figure out how to improve pilot performance.

Is there an accident-prone pilot? A study in 1951 published by Elizabeth Mechem Fuller and Helen B. Baune of the University of Minnesota determined there were injury-prone children. The study was comprised of two separate groups of second grade students. Fifty-five students were considered accident repeaters and 48 students had no accidents. Both groups were from the same school of 600 and their family demographics were similar.

The accident-free group showed a superior knowledge of safety and were considered industrious and cooperative with others but were not considered physically inclined. The accident-repeater group had better gymnastic skills, were considered aggressive and impulsive, demonstrated rebellious behavior when under stress, were poor losers, and liked to be the center of attention. *[Figure 2-2]* One interpretation of this

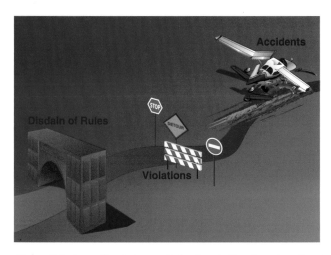

Figure 2-2. *According to human behavior studies, there is a direct correlation between disdain for rules and aircraft accidents.*

data—an adult predisposition to injury stems from childhood behavior and environment—leads to the conclusion that any pilot group should be comprised only of pilots who are safety conscious, industrious, and cooperative. Clearly, this is not only an inaccurate inference, but is impossible to achieve since pilots are drawn from the general population and exhibit all types of personality traits.

Fifty-five years after Fuller-Baune study, Dr. Patrick R. Veillette debated the possibility of an accident prone pilot in his 2006 article "Accident-Prone Pilots," published in Business and Commercial Aviation. Veillette uses the history of "Captain Everyman" to demonstrate how aircraft accidents are caused more by a chain of poor choices than one single poor choice. In the case of Captain Everyman, after a gear-up landing accident, he became involved in another accident while taxiing a Beech 58P Baron out of the ramp. Interrupted by a radio call from the dispatcher, Everyman neglected to complete the fuel cross-feed check before taking off. Everyman, who was flying solo, left the right fuel selector in the cross-feed position. Once aloft and cruising, he noticed a right roll tendency and corrected with aileron trim. He did not realize that both engines were feeding off the left wing's tank, making the wing lighter. *[Figure 2-3]*

After two hours of flight, the right engine quit when Everyman was flying along a deep canyon gorge. While he was trying to troubleshoot the cause of the right engine's failure, the left engine quit. Everyman landed the aircraft on a river sand bar, but it sank into ten feet of water.

Several years later, Everyman was landing a de Havilland Twin Otter when the aircraft veered sharply to the left, departed the runway, and ran into a marsh 375 feet from the runway. The airframe and engines sustained considerable damage. Upon inspecting the wreck, accident investigators found the nosewheel steering tiller in the fully deflected position. Both the after-takeoff and before-landing checklists required the tiller to be placed in the neutral position. Everyman had overlooked this item.

Now, is Everyman accident prone or just unlucky? Skipping details on a checklist appears to be a common theme in the preceding accidents. While most pilots have made similar mistakes, these errors were probably caught prior to a mishap due to extra margin, good warning systems, a sharp copilot, or just good luck. In an attempt to discover what makes a pilot accident prone, the Federal Aviation Administration (FAA) oversaw an extensive research study on the similarities and

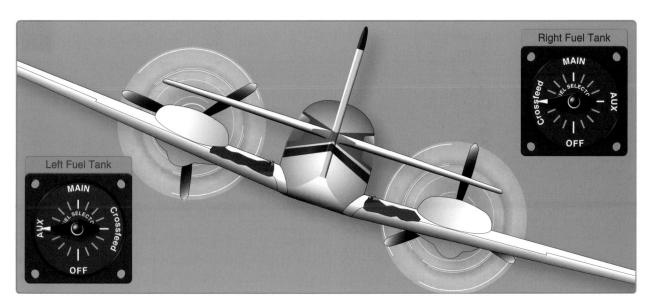

Figure 2-3. *The pilot inadvertently fed both engines from the left fuel tank and failed to determine the problem for the right wing low. His lack of discipline resulted in an accident.*

dissimilarities of pilots who were accident free and those who were not. The project surveyed over 4,000 pilots, half of whom had "clean" records while the other half had been involved in an accident.

Five traits were discovered in pilots prone to having accidents *[Figure 2-4]*:

1. Disdain toward rules

2. High correlation between accidents in their flying records and safety violations in their driving records

3. Frequently falling into the personality category of "thrill and adventure seeking"

4. Impulsive rather than methodical and disciplined in information gathering and in the speed and selection of actions taken

5. Disregard for or underutilization of outside sources of information, including copilots, flight attendants, flight service personnel, flight instructors, and air traffic controllers

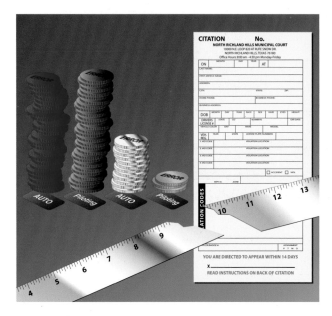

Figure 2-4. *Pilots with hazardous attitudes have a high incident rate of accidents.*

In contrast, the successful pilot possesses the ability to concentrate, manage workloads, monitor, and perform several simultaneous tasks. Some of the latest psychological screenings used in aviation test applicants for their ability to multitask, measuring both accuracy and the individual's ability to focus attention on several subjects simultaneously.

Research has also demonstrated significant links between pilot personality and performance, particularly in the area of crew coordination and resource management. Three distinct

subgroups of flight crew member personalities have been isolated: right stuff, wrong stuff, and no stuff. As the names imply, the right stuff group has the right stuff. This group demonstrates positive levels of achievement motivation and interpersonal behavior. The wrong stuff group has high levels of negative traits, such as being autocratic or dictatorial. The no stuff group scored low on goal seeking and interpersonal behaviors.

These groups became evident in a 1991 study, "Outcomes of Crew Resource Management Training" by Robert L. Helmreich and John A. Wilhelm. During this study a subset of participants reacted negatively to the training–the individuals who seemed to need the training the most were the least receptive. The authors felt that personality factors played a role in this reaction because the ones who reacted negatively were individuals who lacked interpersonal skills and had not been identified as members of the "right stuff" subset. It was surmised that they felt threatened by the emphasis on the importance of communications and human relations skills.

The influence of personality traits can be seen in the way a pilot handles a flight. For example, one pilot may be uncomfortable with approximations and "guesstimates," preferring to use his or her logical, problem-solving skills to maintain control over instrument flight operations. Another pilot, who has strong visual-spatial skills and prefers to scan, may apply various "rules of thumb" during a instrument flight period. The first pilot's personality is reflected in his or her need to be planned and structured. The second type of pilot is more fluid and spontaneous and regards mental calculations as bothersome.

No one ever intends to have an accident and many accidents result from poor judgment. For example, a pilot flying several trips throughout the day grows steadily behind schedule due to late arriving passengers or other delays. Before the last flight of the day, the weather starts to deteriorate, but the pilot thinks one more short flight can be squeezed in. It is only 10 minutes to the next stop. But by the time the cargo is loaded and the flight begun, the pilot cannot see the horizon while flying out over the tundra. The pilot decides to forge on since he told the village agent he was coming and flies into poor visibility. The pilot never reaches the destination and searchers find the aircraft crashed on the tundra.

In this scenario, a chain of events results in the pilot making a poor decision. First, the pilot exerts pressure on himself to complete the flight, and then proceeds into weather conditions that do not allow a change in course. In many such cases, the flight ends in controlled flight into terrain (CFIT).

In a 2005 FAA study, it became apparent that human error associated with GA accidents is multifaceted. Specifically, the analyses revealed that the largest percentage of accidents is associated with skill-based errors, followed by decision errors, violations of the rules and regulations, and perceptual errors. *[Figure 2-5]* The next step will be identifying a variety of interventions targeted at all four error groups. Eliminating human errors is an unrealistic goal since errors are a normal part of human behavior. On the other hand, realizing that many aviation accidents are preventable means designing ways to reduce the consequences of human error. The study of human behavior coupled with pilot training that offsets predictable human error helps achieve that goal.

Chapter Summary

Studies of human behavior help isolate characteristics and behaviors that can lead to poor decision-making by a pilot.

Figure 2-5. *Accident-prone pilots fail to use readily available resources, or they simply do not listen.*

Chapter 3

Identifying Hazards and Mitigating Risk

Introduction

As previously discussed, identifying hazards and associated risk is key to preventing risk and accidents. If a pilot fails to search for risk, it is likely that he or she will neither see it nor appreciate it for what it represents. Unfortunately in aviation, pilots seldom have the opportunity to learn from their small errors in judgment because even small mistakes in aviation are often fatal. In order to identify risk, the use of standard procedures is of great assistance. One guide in the form of a checklist that helps the pilot examine areas of interest in his or her preflight planning is a framework called PAVE. Elements of PAVE are:

Pilot-in-command (PIC)

Aircraft

En**V**ironment

External pressures

Pilot

- A pilot must continually make decisions about competency, condition of health, mental and emotional state, level of fatigue, and many other variables. For example, a pilot may be called early in the morning to make a long flight. If a pilot has had only a few hours of sleep and is concerned that the congestion being experienced could be the onset of a cold, it would be prudent to consider if the flight could be accomplished safely.

- A pilot had only 4 hours of s... being asked by the boss to f... city 750 miles away. The rep... marginal and not expected to... assessing fitness as a pilot, ... would not be wise to make th...

Aircraft

- A Pilot will frequently base decisions on the evaluat... airplane, such as performance, equipment, or airwo...

- During a preflight, a pilot noticed a small amount of... from the bottom of the cowling. Although the quant... seemed insignificant at the time, the pilot decided t... takeoff and have a mechanic check the source of th... The pilot's good judgment was confirmed when the... found that one of the oil cooler hose fittings was lo...

EnVironment

- This encompasses many elements not pilot or ai... It can include such factors as weather, air traffic control, navigational aids (NAVAIDS), terrain, takeoff and landing areas, and surrounding obstacles. Weather is one element that can change drastically over time and distance.

- A pilot was landing a small airplane just after a heavy jet had departed a parallel runway. The pilot assumed that wake turbulence would not be a problem since landings had been performed under similar circumstances. Due to a combination of prevailing winds and wake turbulence from the ... the landing ... de a hard ... e an error ... ight

External Pressures

- The interaction between the pilot, airplane, and the environment is greatly influenced by the purpose of each flight operation. The pilot must evaluate the three previous areas to decide on the desirability of undertaking or continuing the flight as planned. It is worth asking why the flight is being made, how critical is it to maintain the schedule, and is the trip worth the risks?

- On a ferry flight to deliver an airplane from the factory, in marginal weather conditions, the pilot calculated the groundspeed and determined that the airplane would arrive at the destination with only 10 minutes of fuel remaining. The pilot was determined to keep on schedule by trying to "stretch" the fuel supply instead of landing to refuel. After landing with low fuel state, the pilot realized that this could have easily resulted in an emergency landing in deteriorating weather conditions. This was a chance that was not worth taking to keep the planned schedule.

NOT HERE
HERE

Gusting to 28 k...

✓ **I'M SAFE CHECKLIST**

Stress—Do I have any symptoms?

Medication—Have I been taking prescription or over-the-counter drugs?

Stress—Am I under psychological pressure from the job? Worried about financial matters, health problems, or family discord?

Alcohol—Have I been drinking within 8 hours? within 24 hours?

Fatigue—Am I tired and not adequately rested?

Eating—Am I adequately nourished?

Using PAVE helps to identify risk before departure and assists the pilot's decision-making process. *[Figure 3-1]*

With the PAVE checklist, pilots have a simple way to remember each category to examine for risk prior to each flight. Once a pilot identifies the risks of a flight, he or she needs to decide whether the risk or combination of risks can be managed safely and successfully. If not, make the decision to cancel the flight. If the pilot decides to continue with the flight, he or she should develop strategies to mitigate the risks. One way a pilot can control the risks is to set personal minimums for items in each risk category. These are limits

Pilot

- A pilot must continually make decisions about competency, condition of health, mental and emotional state, level of fatigue, and many other variables. For example, a pilot may be called early in the morning to make a long flight. If a pilot has had only a few hours of sleep and is concerned that the sinus congestion being experienced could be the onset of a cold, it would be prudent to consider if the flight could be accomplished safely.

- A pilot had only 4 hours of sleep the night before being asked by the boss to fly to a meeting in a city 750 miles away. The reported weather was marginal and not expected to improve. After assessing fitness as a pilot, it was decided that it would not be wise to make the flight. The boss was initially unhappy, but was later convinced by the pilot that the risks involved were unacceptable.

Aircraft

- A pilot frequently bases decisions on evaluation of the airplane, such as performance, equipment, or airworthiness.

- During a preflight, a pilot noticed a small amount of oil dripping from the bottom of the cowling. Although the quantity of oil seemed insignificant at the time, the pilot decided to delay the takeoff and have a mechanic check the source of the oil. The pilot's good judgment was confirmed when the mechanic found that one of the oil cooler hose fittings was loose.

En**V**ironment

- The environment encompasses many elements that are not pilot or airplane related, including such factors as weather, air traffic control (ATC), navigational aids (NAVAIDS), terrain, takeoff and landing areas, and surrounding obstacles. Weather is one element that can change drastically over time and distance.

- A pilot was landing a small airplane just after a heavy jet had departed a parallel runway. The pilot assumed that wake turbulence would not be a problem since landings had been performed under similar circumstances. Due to a combination of prevailing winds and wake turbulence from the heavy jet drifting across the landing runway, the airplane made a hard landing. The pilot made an error when assessing the flight environment.

External Pressures

- The interaction between the pilot, airplane, and the environment is greatly influenced by the purpose of each flight operation. The pilot must evaluate the three previous areas to decide on the desirability of undertaking or continuing the flight as planned. It is worth asking why the flight is being made, how critical it is to maintain the schedule, and if the trip is worth the risks.

- On a ferry flight to deliver an airplane from the factory, the pilot calculated the groundspeed and determined he would arrive at the destination with only 10 minutes of fuel remaining. A check of the weather revealed he would be flying into marginal weather conditions. By asking himself whether it was more critical to maintain the schedule or to arrive with an intact aircraft, the pilot decided to schedule a refuel stop even though it would mean he would not be able to keep to the schedule. He chose not to "stretch" the fuel supply in marginal weather conditions which could have resulted in an emergency landing.

Figure 3-1. *The PAVE checklist.*

unique to that individual pilot's current level of experience and proficiency.

One of the most important concepts that safe pilots understand is the difference between what is "legal" in terms of the regulations, and what is "smart" or "safe" in terms of pilot experience and proficiency.

P = Pilot in command

The pilot in command (PIC) *[Figure 3-2]* is one of the risk factors in a flight. The pilot must ask, "Am I ready for this trip?" in terms of experience, currency, physical, and emotional condition.

The Pilot's Health

One of the best ways pilots can mitigate risk is a self-evaluation to ensure they are in good health. A standardized method used in evaluating health employs the IMSAFE checklist. *[Figure 3-3]* It can easily and effectively be used to determine physical and mental readiness for flying and provides a good overall assessment of the pilot's well being.

1. Illness—Am I sick? Illness is an obvious pilot risk.

2. Medication—Am I taking any medicines that might affect my judgment or make me drowsy?

3. Stress—Am I under psychological pressure from the job? Do I have money, health, or family problems? Stress causes concentration and performance problems.

Figure 3-2. *The highest risk for the pilot is self, and requires special introspective analysis.*

✓ I'M SAFE CHECKLIST

Illness—Do I have any symptoms?

Medication—Have I been taking prescription or over-the-counter drugs?

Stress—Am I under psychological pressure from the job? Worried about financial matters, health problems, or family discord?

Alcohol—Have I been drinking within 8 hours? Within 24 hours?

Fatigue—Am I tired and not adequately rested?

Emotion—Am I emotionally upset?

Figure 3-3. IMSAFE checklist.

Stressors
Environmental Conditions associated with the environment, such as temperature and humidity extremes, noise, vibration, and lack of oxygen.
Physiological Stress Physical conditions, such as fatigue, lack of physical fitness, sleep loss, missed meals (leading to low blood sugar levels), and illness.
Psychological Stress Social or emotional factors, such as a death in the family, a divorce, a sick child, or a demotion at work. This type of stress may also be related to mental workload, such as analyzing a problem, navigating an aircraft, or making decisions.

Figure 3-4. System stressors have a profound impact, especially during periods of high workload.

While the regulations list medical conditions that require grounding, stress is not among them. The pilot should consider the effects of stress on performance.

4. Alcohol—Have I been drinking within 8 hours? Within 24 hours? As little as one ounce of liquor, one bottle of beer, or four ounces of wine can impair flying skills. Alcohol also renders a pilot more susceptible to disorientation and hypoxia.

5. Fatigue—Am I tired and not adequately rested? Fatigue continues to be one of the most insidious hazards to flight safety, as it may not be apparent to a pilot until serious errors are made.

6. Emotion—Have I experienced any emotionally upsetting event?

Stress Management

Everyone is stressed to some degree almost all of the time. A certain amount of stress is good since it keeps a person alert and prevents complacency. Effects of stress are cumulative and, if the pilot does not cope with them in an appropriate way, they can eventually add up to an intolerable burden. Performance generally increases with the onset of stress, peaks, and then begins to fall off rapidly as stress levels exceed a person's ability to cope. The ability to make effective decisions during flight can be impaired by stress. There are two categories of stress—acute and chronic. These are both explained in Chapter 16, Aeromedical Factors, of the Pilot's Handbook of Aeronautical Knowledge. Factors referred to as stressors can affect decision-making skills and increase a pilot's risk of error in the flight deck. *[Figure 3-4].*

For instance, imagine a cabin door that suddenly opens in flight on a Bonanza climbing through 1,500 feet on a clear sunny day? It may startle the pilot, but the stress would wane when it became apparent that the situation was not a serious hazard. Yet, if the cabin door opened in instrument meteorological conditions (IMC), the stress level would be much higher despite little difference between the two scenarios. Therefore, one can conclude that our perception of problems (and the stress they create) is related to the environment in which the problems occur.

Another example is that mechanical problems always seem greater at night, a situation that all pilots have experienced. The key to stress management is to stop, think, and analyze before jumping to a conclusion. There is usually time to think before drawing conclusions.

There are several techniques to help manage the accumulation of life stress, and prevent stress overload. For example, to help reduce stress levels, set aside time for relaxation each day or maintain a program of physical fitness. To prevent stress overload, learn to manage time more effectively to avoid pressures imposed by getting behind schedule and not meeting deadlines.

A = Aircraft

What about the aircraft? What limitations will the aircraft impose upon the trip? Ask yourself the following questions:

* Is this the right aircraft for the flight?

* Am I familiar with and current in this aircraft? Aircraft performance figures and the aircraft flight manual (AFM) are based on a new aircraft flown by a professional test pilot, factors to keep in mind while assessing personal and aircraft performance.

* Is this aircraft equipped for the flight? Instruments? Lights? Are the navigation and communication equipment adequate?

- Can this aircraft use the runways available for the trip with an adequate margin of safety under the conditions to be flown? For instance, consider an AFM for an aircraft that indicates a maximum demonstrated crosswind component of 15 knots. What does this mean to a pilot? This is the maximum crosswind that the manufacturer's test pilot demonstrated in the aircraft's certification. *[Figure 3-5]*

- Can this aircraft carry the planned load?

- Can this aircraft operate with the equipment installed?

- Does this aircraft have sufficient fuel capacity, with reserves, for trip legs planned?

- Is the fuel quantity correct? Did I check? (Remember that most aircraft are manufactured to a standard that requires the fuel indicator be accurate when the fuel quantity is full.)

Using the PAVE checklist would help elevate risks that a pilot may face while preparing and conducting a flight. In the case presented in *Figure 3-5*, the pilot disregarded the risk, failed to properly evaluate its impact upon the mission, or incorrectly perceived the hazard and had an inaccurate perception of his skills and abilities.

At 1030, Cessna 150M veered off the runway and collided with a ditch during a crosswind landing. The private pilot, the sole occupant, sustained minor injuries; the airplane sustained substantial damage. The pilot stated in a written report that he configured the airplane for a straight in approach to runway 27. After touchdown, the airplane veered to the left and departed the runway. The airplane continued through an adjacent field and collided with a ditch. The airplane sustained a buckled firewall and a bent left wing spar. The closest official weather observation was 8 nautical miles (NM) east of the accident site. An aviation routine weather report (METAR) was issued at 0954. It stated: winds from 360 degrees at 19 knots gusting to 28 knots; visibility 10 miles; skies 25,000 feet scattered; temperature 25 °C; dew point 2 °C; altimeter 30.04" Hg.

Figure 3-5. *Considering the crosswind component.*

V = Environment

Weather
Weather is a major environmental consideration. As pilots set their own personal minimums, they should evaluate the weather for a particular flight by considering the following:

- What are the current ceiling and visibility? In mountainous terrain, consider having higher minimums for ceiling and visibility, particularly if the terrain is unfamiliar.

- Consider the possibility that the weather may be different from forecast. Have alternative plans and be ready and willing to divert should an unexpected change occur.

- Consider the winds at the airports being used and the strength of the crosswind component. *[Figure 3-5]*

- If flying in mountainous terrain, consider whether there are strong winds aloft. Strong winds in mountainous terrain can cause severe turbulence and downdrafts and be very hazardous for aircraft even when there is no other significant weather.

- Are there any thunderstorms present or forecast?

- If there are clouds, is there any icing, current or forecast? What is the temperature-dew point spread and the current temperature at altitude? Can descent be made safely all along the route?

- If icing conditions are encountered, is the pilot experienced at operating the aircraft's deicing or anti-icing equipment? Is this equipment in good condition and functional? For what icing conditions is the aircraft rated, if any?

Terrain
Evaluation of terrain is another important component of analyzing the flight environment.

- To avoid terrain and obstacles, especially at night or in low visibility, determine safe altitudes in advance by using the altitudes shown on visual flight rules (VFR) and instrument flight rules (IFR) charts during preflight planning.

- Use maximum elevation figures (MEF) *[Figure 3-6]* and other easily obtainable data to minimize chances of an inflight collision with terrain or obstacles.

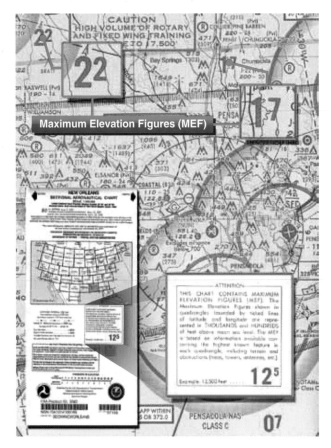

Figure 3-6. *The pilot can easily assess elevations at a glance by simply comparing the intended altitude to the minimum elevation figures (MEFs) depicted on all VFR sectional charts. The MEFs are one of the best sources of elevation information and can be used during both the planning and flight phases.*

Airport

- What lights are available at the destination and alternate airports (e.g., visual approach slope indicator (VASI), precision approach path indicator (PAPI) or instrument landing system (ILS), glideslope guidance)? *[Figure 3-7]* Is the terminal airport equipped with them? Are they working? Will the pilot need to use the radio to activate the airport lights?

- Check the Notices to Airmen (NOTAMS) for closed runways or airports. Look for runway or beacon lights out, nearby towers, etc.

- Choose the flight route wisely. An engine failure gives the nearby airports supreme importance.

- Are there shorter or obstructed fields at the destination and/or alternate airports?

Airspace

- If the trip is over remote areas, are appropriate clothing, water, and survival gear onboard in the event of a forced landing?

- If the trip includes flying over water or unpopulated areas with the chance of losing visual reference to the horizon, the pilot must be prepared to fly IFR.

- Check the airspace and any temporary flight restrictions (TFRs) along the route of flight.

Nighttime

Night flying requires special consideration.

- If the trip includes flying at night over water or unpopulated areas with the chance of losing visual

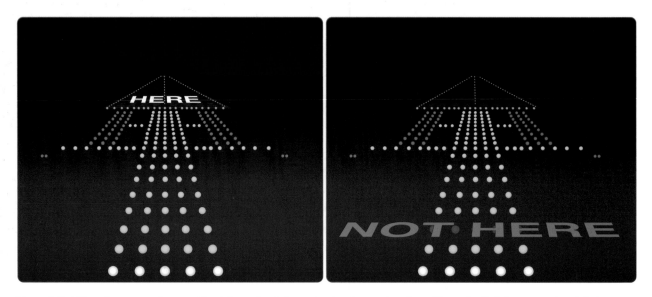

Figure 3-7. *Although runways that provide plain-spoken information (as shown above) would require little interpretation, it is important to understand and interpret runway indicators used in the aviation environment.*

reference to the horizon, the pilot must be prepared to fly IFR.

- Will the flight conditions allow a safe emergency landing at night?

- Preflight all aircraft lights, interior and exterior, for a night flight. Carry at least two flashlights—one for exterior preflight and a smaller one that can be dimmed and kept nearby. *[Figure 3-8]*

Figure 3-8. A *chemical stick is useful to carry onboard the aircraft at night. It comes in various colors, intensities, and durations, and it provides ample illumination within the flight deck. This does not replace the regulatory requirement of carrying flashlights.*

The human eye will see nothing outside that is dimmer than the flight deck lighting. Always fly at night with the interior lights as dim as possible. As the flight progesses and the eyes adjust to the darkness, usually the interior lights can be dimmed further, aiding the outside vision. If the interior lights will not dim, that would increase the risk factors by restricting the pilot's outside vision—probably not the time for a night flight.

Visual Illusions

Although weather, terrain, airport conditions, and night versus daylight flying each produce unique challenges, together these factors conspire against a pilot's senses. It is important to understand that unwittingly these factors can create visual illusions and cause spatial disorientation producing challenges the pilot did not anticipate. *[Figure 3-9]* Even the best trained pilots sometimes fail to recognize a problem until it is too late to complete a flight safely.

An accident involving a Piper PA-32 and an airline transport pilot illustrates how visual illusions can create problems that lead to an accident. In this case, the aircraft collided with terrain during a landing. The sole occupant of the airplane was an airline transport pilot who was not injured. The airplane owned and operated by the pilot, sustained substantial damage. The personal transportation flight was being operated in visual meteorological conditions (VMC) in mid-afternoon. Although it was not snowing, there was snow on the ground.

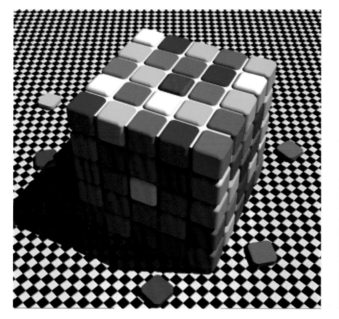

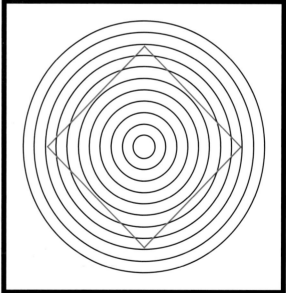

Figure 3-9. *Visual illusions are easy to see when shown in the examples above. The illusion on the left represents how the brain processes color. The "brown" square on top and the "orange" square on the side are actually the same color. The illusion on the right appears to have red lines that curve; however, they are straight. These illusions are representative of things we see in everyday life, except we do not see them as they really are until it is sometimes too late. Understanding that visual illusions exist is a prime ingredient to being better prepared to cope with risk.*

Originally on an IFR flight plan, the pilot canceled his IFR clearance when he had the airport in sight. According to the pilot, he was familiar with the airport, having landed there repeatedly in the past. However, it had just snowed, leaving a thin layer of snow and mixed ice on the runway. The pilot in this case allowed his familiarity with the airport coupled with his flight experience give him a false sense of confidence. As a result, he failed to realistically assess the potential snow and ice hazard on the runway—an assessment overshadowed by his own self-assurance exacerbated by his familiarity and experience.

On the day of the accident, the runway was covered with one inch of snow and ice. Previously plowed snow lined the runway. Although he had not landed on a snow-covered runway in 10 years, the pilot felt his knowledge of the runway environment and familiarity with the airfield would compensate for this lack of currency in landing in these types of conditions. During the final approach, the visual cues normally available to a pilot were not present. That is, the snow-covered terrain presented problems for the pilot in ascertaining proper depth of field, recognized as a visual illusion. When he landed, his normally available lateral visual cues were obscured by the snow, causing him to come in at a higher altitude than he normally would have. Disoriented by the snow and lacking knowledge on how to adapt properly to these conditions, he was unable to determine his position relative to the runway centerline and landed left of the intended point. By focusing his attention on the snow banks, he drifted further toward the edge of the runway causing one of the airplane's main gears to miss the runway surface.

The risk at hand could be addressed in the following manner. Does landing on snow and ice require any special skills? Do you have these skills? Are you current in using these skills? If landing in ice and snow requires special airmanship skills that transcend normal pilotage and you do not have that skill or you are no longer experienced in this situation, then the risk is increased and you need to recognize that just because you are a pilot does not mean you are proficient at doing all of the maneuvers you are legally qualified to perform. Examine seaplane ratings, mountain training, and tail-wheel proficiency. This proficiency starts to wane the moment a pilot stops performing maneuvers requiring these skills.

Immediately after touching down, the wheel that was off the edge of the runway hit a snow-covered mound of previously plowed snow. The impact threw the airplane sideways and it collided with more of the previously plowed snow. During this sequence, all three landing gear struts collapsed and the underside of the airplane sustained considerable structural damage.

What could this pilot have done to prevent this accident from happening? In addition to maintaining currency in landing on a snow-covered runway, he could have prevented this accident by choosing an alternate airport that had a cleared runway. He could have taken another pilot, junior or senior to his overall experience who has landed in similar conditions recently. Certainly he could have been better prepared. He could have read about landing in these conditions and better prepared himself for landing on snow and ice. He could have planned. Before landing on snow-covered terrain, a pilot needs to understand how to accomplish the landing since the techniques are not the same as those for landing on a clear, dry runway. In this example, the pilot applied the same methods of ascertaining depth perception as normally used if the terrain were not blanketed in snow.

In this case, the basic underlying problem was the pilot's failure to prepare for the conditions. He knew the challenge that faced him, and he had the assets to prepare himself better, yet he did not. In reality, the hazard in this case is not just the snow or the challenges it presented, but the pilot himself in being overly confident and even complacent to his responsibilities. Had this aircraft been carrying passengers and had the accident occurred under slightly different conditions, the end result could have been tragic.

The first and key step in preparing for a new situation is to recognize that one may not have the required skill set—the step of recognzing personal limitations. The next step is acquiring that skill set. A pilot who has never landed on snow, or one whose skills have eroded from lack of recent practice, can do the following to acquire or renew the skill set necessary for a successful landing in snow conditions:

1. Review reference materials to reinforce and increase knowledge about visual illusions and their causes:

   • Aeronautical Information Manual (AIM) Chapter 8, Medical Facts for Pilots

   • Pilot's Handbook of Aeronautical Knowledge, Chapter 15, Navigation

   • Advisory Circular (AC) 60-4, Spatial Disorientation

   • AC 90-48, Pilot's Role in Collision Avoidance

2. Fly with an instructor pilot or other PIC who has had significant experience in landing on snow.

3. Participate in a training designed specifically for landing in unusual places and environments. Many pilots attend classes on mountain flying in which they learn techniques to use in the absence of standard visual cues.

E = External Pressures

External pressures are influences external to the flight that create a sense of pressure to complete a flight—often at the expense of safety. Factors that can be external pressures include the following:

- Someone waiting at the airport for the flight's arrival

- A passenger the pilot does not want to disappoint

- The desire to demonstrate pilot qualifications

- The desire to impress someone (Probably the two most dangerous words in aviation are "Watch this!")

- Desire to satisfy a specific personal goal ("get-home-itis," "get-there-itis," and "let's-go-itis")

- A pilot's general goal-completion orientation

- The emotional pressure associated with acknowledging that skill and experience levels may be lower than a pilot would like them to be. (Pride can be a powerful external factor.)

The following accident offers an example of how external pressures influence a pilot. Two pilots were giving helicopter demonstrations at an air show. The first pilot demonstrated a barrel roll in front of the stands. Not to be outdone, the second pilot (with passengers) decided to execute a hammerhead type maneuver. Flying past the stands at 90 knots, the pilot pulled the helicopter into a steep climb that ended at about 200 feet. When the speed dissipated to near zero, he rolled back to the ground in a nose-low attitude to regain airspeed with the obvious intention of pulling the aircraft out of the dive near the ground. An error in judgment led to the pilot being unable to pull the helicopter out of the dive. The helicopter struck the ground, killing all onboard.

The desire to impress someone can be a powerful external pressure, especially when coupled with the internal pressure of pride. Perhaps the pilot decided to perform a maneuver not in his training profile, or one in which he had not demonstrated proficiency. It appears there was nothing in this pilot's experiences to help him effectively access the high risk of this maneuver in an aircraft loaded with passengers. It is not uncommon to see people motivated by external pressures who are also driven internally by their own attitude.

Management of external pressure is the single most important key to risk management because it is the one risk factor category that can cause a pilot to ignore all other risk factors. External pressures place time-related pressure on the pilot and figure into a majority of accidents.

Helicopter Emergency Medical Service (HEMS) operations, unique due to the emergency nature of the mission, are an example of how external pressures influence pilots. Emergency medical services (EMS) pilots often ferry critically ill patients, and the pilot is driven by goal completion. In order to reduce the effect of this pressure, many EMS operators do not to notify the EMS pilot of the prospective patient's condition, but merely confine the location of the patient pickup and restrict the pilot's decision-making role to the response to the question "Can the pickup and transportation to the medical care center be made safely?" Risking three or four lives in an attempt to save one life is not a safe practice.

The use of personal standard operating procedures (SOPs) is one way to manage external pressures. The goal is to supply a release for the external pressures of a flight. These procedures include, but are not limited to:

- Allow time on a trip for an extra fuel stop or to make an unexpected landing because of weather.

- Have alternate plans for a late arrival or make backup airline reservations for must-be-there trips.

- For really important trips, plan to leave early enough so that there would still be time to drive to the destination.

- Advise those who are waiting at the destination that the arrival may be delayed. Know how to notify them when delays are encountered.

- Manage passenger expectations. Ensure passengers know that they might not arrive on a firm schedule, and if they must arrive by a certain time, they should make alternative plans.

- Eliminate pressure to return home, even on a casual day flight, by carrying a small overnight kit containing prescriptions, contact lens solutions, toiletries, or other necessities on every flight.

The key to managing external pressure is to be ready for and accept delays. Remember that people get delayed when traveling on airlines, driving a car, or taking a bus. The pilot's goal is to manage risk, not increase it.

Chapter Summary

Risk can be identified and mitigated by using checklists such as PAVE and IMSAFE. Accident data offers the opportunity to explain how pilots can use risk management to increase the safety of a flight.

Chapter 4
Assessing Risk

Introduction

Assessment of risk is an important component of good risk management, but before a pilot can begin to assess risk, he or she must first perceive the hazard and attendant risk(s). In aviation, experience, training, and education help a pilot learn how to spot hazards quickly and accurately. Once a hazard is identified, determining the probability and severity of an accident (level of risk associated with it) becomes the next step. For example, the hazard of a nick in the propeller poses a risk only if the airplane is flown. If the damaged prop is exposed to the constant vibration of normal engine operation, there is a high risk that it could fracture and cause catastrophic damage to the engine and/or airframe and the passengers.

Risk Assessment Matrix

Likelihood	Severity			
	Catastrophic	Critical	Marginal	Negligible
Probable	High	High	Serious	
Occasional	High	Serious		
Remote	Serious	Medium		
Improbable				Low

RISK ASSESSMENT

SLEEP
Did not sleep well or less than 8 hours — 2
Slept well — 0

HOW DO YOU FEEL?
1. Have a cold or ill — 4
2. Feel great — 0
3. Feel a bit off — 2

WEATHER AT TERMINATION
1. Greater than 5 miles visibility and 3,000 feet ceilings — 1
2. At least 3 miles visibility and 1,000 feet ceilings, but less than 3,000 feet ceilings and 5 miles visibility — 3
3. IMC conditions — 4

Column total

HOW IS THE DAY GOING?

IS THE FLIGHT
1. Day?
2. Night?

PLANNING

Low Risk — Not Complex Flight — 0 — 10 — Exercise Caution — 20 — Area of Concern — 30 — Endangerment — LEFT COLUMN TOTAL

Every flight has hazards and some level of risk associated with it. Pilots must recognize hazards to understand the risk they present. Knowing that risk is dynamic, one must look at the cumulative effect of multiple hazards facing us. It is critical that pilots are able to:

- Differentiate, in advance, between a low-risk flight and a high-risk flight.

- Establish a review process and develop risk mitigation strategies to address flights throughout that range.

For the pilot who is part of a flight crew, input from various responsible individuals cancels out any personal bias or skewed judgment during preflight planning and the discussion of weather parameters. The single pilot does not have the advantage of this oversight. If the single pilot does not comprehend or perceive the risk, he or she will make no attempt to mitigate it. The single pilot who has no other crewmember for consultation must be aware of hazardous conditions that can lead to an accident. Therefore, he or she has a greater vulnerability than a pilot with a full crew.

Assessing risk is not always easy, especially when it involves personal quality control. For example, if a pilot who has been awake for 16 hours and logged over 8 hours of flight time is asked to continue flying, he or she will generally agree to continue flying. Pilots often discount the fatigue factor because they are goal oriented and tend to deny personal limitations when asked to accept a flight. This tendency is exemplified by pilots of helicopter emergency medical services (EMS) who, more than other pilot groups, may make flight decisions based upon the patient's welfare rather than the pilot's personal limitations. These pilots weigh intangible factors such as the patient's condition and fail to quantify actual hazards appropriately, such as fatigue or weather, when making flight decisions.

Examining National Transportation Safety Board (NTSB) reports and other accident research can help a pilot learn to assess risk more effectively. For example, the accident rate during night visual flight rules (VFR) decreases by nearly 50 percent once a pilot obtains 100 hours, and continues to decrease until the 1,000 hour level. The data suggest that for the first 500 hours, pilots flying VFR at night might want to establish higher personal limitations than are required by the regulation and, if applicable, become better skilled at flying under instrument conditions.

Several risk assessment models are available to assist the pilot in determining his or her risk before departing on a flight. The models, all taking slightly different approaches, seek the common goal of assessing risk in an objective manner.

Quantifying Risk Using a Risk Matrix

The most basic tool is the risk matrix. *[Figure 4-1]* It assesses two items: the likelihood of an event occurring and the consequence of that event.

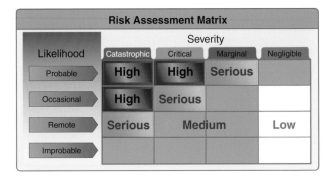

Figure 4-1. *A sample risk assessment matrix a pilot can use to differentiate between low-risk and high-risk flights.*

Likelihood of an Event

Likelihood is nothing more than taking a situation and determining the probability of its occurrence. It is rated as probable, occasional, remote, or improbable. For example, a pilot is flying from point A to point B (50 miles) in marginal visual flight rules (MVFR) conditions. The likelihood of encountering potential instrument meteorological conditions (IMC) is the first question the pilot needs to answer. The experiences of other pilots coupled with the forecast might cause the pilot to assign "occasional" to determine the probability of encountering IMC.

The following are guidelines for making assignments.

- Probable—an event will occur several times.

- Occasional—an event will probably occur sometime.

- Remote—an event is unlikely to occur, but is possible.

- Improbable—an event is highly unlikely to occur.

Severity of an Event

The other item in the matrix is the severity or consequence of a pilot's action(s). It can relate to injury and/or damage. If the individual in the example above is not an instrument flight rules (IFR) pilot, what are the consequences of encountering inadvertent IMC conditions? In this case, because the pilot is not IFR rated, the consequences are potentially catastrophic. The following are guidelines for this assignment.

- Catastrophic—results in fatalities, total loss

- Critical—severe injury, major damage

- Marginal—minor injury, minor damage

- Negligible—less than minor injury, less than minor system damage

Simply connecting the two factors as shown in *Figure 4-1* indicates the risk is high and the pilot must not fly, or fly only after finding ways to mitigate, eliminate, or control the risk.

Although the matrix in *Figure 4-1* provides a general viewpoint of a generic situation, a more comprehensive program can be made that is tailored to a pilot's flying. *[Figure 4-2]* This program includes a wide array of aviation related activities specific to the pilot and assesses health, fatigue, weather, capabilities, etc. The scores are added and the overall score falls into various ranges, with the range representative of actions that a pilot imposes upon himself or herself.

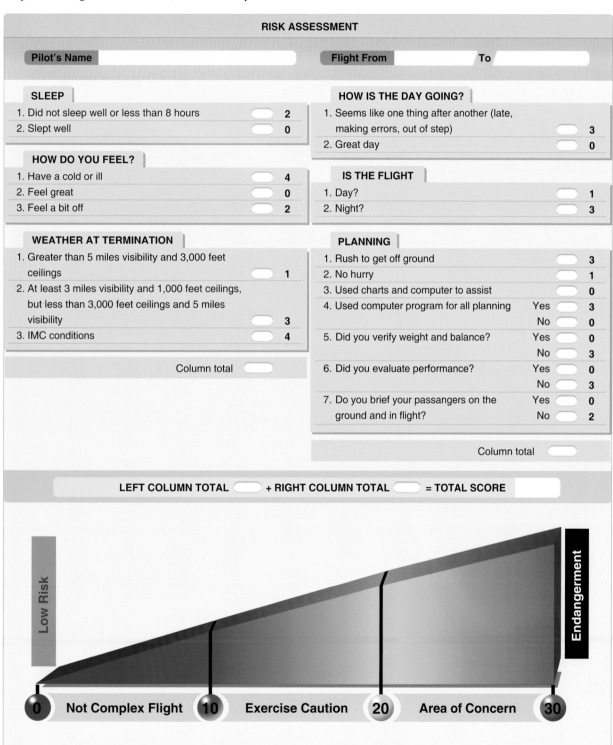

Figure 4-2. *Example of a more comprehensive risk assessment program.*

Mitigating Risk

Risk assessment is only part of the equation. After determining the level of risk, the pilot needs to mitigate the risk. For example, the VFR pilot flying from point A to point B (50 miles) in marginal flight conditions has several ways to reduce risk:

- Wait for the weather to improve to good VFR conditions.

- Take a pilot who is more experienced or who is certified as an instrument flight rules (IFR) pilot.

- Delay the flight.

- Cancel the flight.

- Drive.

Chapter Summary

The pilot can assess risk by using risk assessment models that quantify the risk by assessing the likelihood of an event occurring and the consequences of that event.

Go to www.FAA.gov for a downloadable risk assessment program to use.

Chapter 5

Aeronautical Decision-Making: A Basic Staple

Introduction

Aeronautical decision-making (ADM) is a cornerstone in managing risk. ADM provides a structured framework utilizing known processes and applying recognized pathways, which individually and collectively have a positive effect on exposure to hazards. This is not achieved by reducing the hazard itself, but by helping the pilot recognize hazards that need attention.

ADM is a systematic approach to the mental process used by pilots to consistently determine the best course of action in response to a given set of circumstances. It is what a pilot intends to do based on the latest information he or she has.

The importance of learning and understanding effective ADM skills cannot be overemphasized. While progress is continually being made in the advancement of pilot training methods, aircraft equipment and systems, and services for pilots, accidents still occur. Despite all the changes in technology to improve flight safety, one factor remains the same: the human factor, which leads to errors. It is estimated that approximately 80 percent of all aviation accidents are related to human factors, and the vast majority of these accidents occur during landing (24.1 percent) and takeoff (23.4 percent).

ADM helps reduce risk. To understand ADM is to understand also how personal attitudes can influence decision-making and how those attitudes can be modified to enhance safety in the flight deck. It is important to understand the factors that cause humans to make decisions and how the decision-making process not only works, but also can be improved.

This chapter focuses on helping the pilot improve his or her ADM skills with the goal of mitigating the risk factors associated with flight. Advisory Circular (AC) 60-22, Aeronautical Decision Making, provides background references, definitions, and other pertinent information about ADM training in the general aviation (GA) environment. *[Figure 5-1]*

History of ADM

For over 25 years, the importance of good pilot judgment, or ADM, has been recognized as critical to the safe operation of aircraft, as well as accident avoidance. Research in this area prompted the Federal Aviation Administration (FAA) to produce training directed at improving the decision-making of pilots and led to current FAA regulations that require that decision-making be taught as part of the pilot training curriculum. ADM research, development, and testing culminated in 1987 with the publication of six manuals oriented to the decision-making needs of variously rated pilots. These manuals provided multifaceted materials designed to reduce the number of decision-related accidents. The effectiveness of these materials was validated in independent studies where student pilots received such training in conjunction with the standard flying curriculum. When tested, the pilots who had received ADM training made fewer in-flight errors than those who had not received ADM training. The differences were statistically significant and ranged from about 10 to 50 percent fewer judgment errors. In the operational environment, an operator flying about 400,000 hours annually demonstrated a 54 percent reduction in accident rate after using these materials for recurrency training.

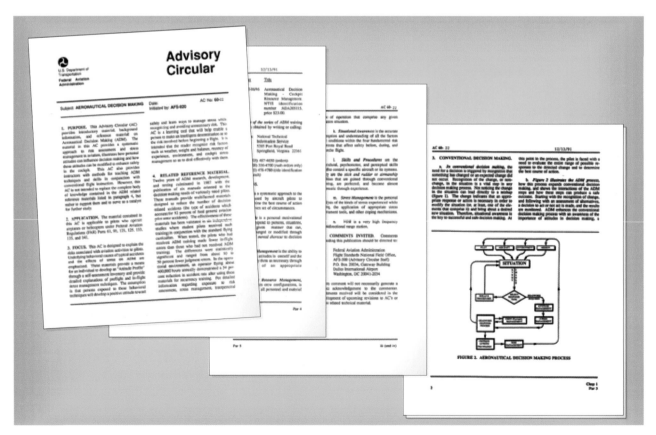

Figure 5-1. *Advisory Circular (AC) 60-22, Aeronautical Decision Making, carries a wealth of information for the pilot to learn.*

Contrary to popular belief, good judgment can be taught. Tradition held that good judgment was a natural by-product of experience, and as pilots continued to log accident-free flight hours, a corresponding increase of good judgment was assumed. Building upon the foundation of conventional decision-making, ADM enhances the process to decrease the probability of human error and increase the probability of a safe flight. ADM provides a structured, systematic approach to analyzing changes that occur during a flight and how these changes might affect a flight's safe outcome. The ADM process addresses all aspects of decision-making in the flight deck and identifies the steps involved in good decision-making.

Steps for good decision-making are:

1. Identifying personal attitudes hazardous to safe flight.

2. Learning behavior modification techniques.

3. Learning how to recognize and cope with stress.

4. Developing risk assessment skills.

5. Using all resources.

6. Evaluating the effectiveness of one's ADM skills.

ADM results in helping to manage risk. When a pilot follows good decision-making practices, the inherent risk in a flight is reduced or even eliminated. The ability to make good decisions is based upon direct or indirect experience and education.

Consider automobile seat belt use. In just two decades, seat belt use has become the norm, placing those who do not wear seat belts outside the norm, but this group may learn to wear a seat belt by either direct or indirect experience. For example, a driver learns through direct experience about the value of wearing a seat belt when he or she is involved in a car accident that leads to a personal injury. An indirect learning experience occurs when a loved one is injured during a car accident because he or she failed to wear a seat belt.

While poor decision-making in everyday life does not always lead to tragedy, the margin for error in aviation is narrow. Since ADM enhances management of an aeronautical environment, all pilots should become familiar with and employ ADM.

Analytical Decision-Making

Analytical decision-making is a form of decision-making that takes both time and evaluation of options. A form of this type of decision-making is based upon the acronym "DECIDE." It provides a six-step process for the pilot to logically make good aeronautical decisions. For example, a pilot who flew from Houston, Texas to Jacksonville, Florida in a Merlin failed to use the decision-making process correctly and to his advantage. Noteworthy about this example is how easily pilots are swayed from taking best courses of action when convenient courses are interpreted as being in our best interest.

Detect a change or hazard. In the case at hand, the pilot was running late after conducting business meetings early in the morning. He and his family departed one hour later than expected. In this case, one would assess the late departure for impact to include the need to amend the arrival time. However, if the pilot is impetuous, these circumstances translate into a hazard. Because this pilot was in a hurry, he did not assess for impact and, as a result, did not amend the arrival time. Key in any decision-making is detecting the situation and its subtleties as a hazard; otherwise, no action is taken by the pilot. It is often the case that the pilot fails to see the evolving hazard. On the other hand, a pilot who does see and understand the hazard, yet makes a decision to ignore it, does not benefit from a decision-making process; the issue is not understanding decision-making, but one of attitude.

Estimate the need to counter or react to the change. As the pilot progressed to the destination, it became apparent that the destination weather (at Craig Field in Jacksonville) was forecast as below approach minimums (due to fog) at the time of arrival. However, weather at an alternative airport just 40 miles away was visual flight rules (VFR). At this time, the pilot should have assessed several factors to include the probability of making a successful approach and landing at Craig versus using an alternative field. In one case, the approach is certainly challenging, but it is an approach at the intended destination. The other location (unencumbered by weather) is inconvenient to the personnel waiting on the ground, requiring that they drive 40 miles to meet the pilot and his family.

Choose a desirable outcome for the flight. Selecting a desirable outcome requires objectivity, and this is when pilots make grave errors. Instead of selecting the course of outcome with consideration to challenges of airmanship, pilots typically select an outcome that is convenient for both themselves and others. And without other onboard or external input, the choice is not only flawed but also reinforced by their own rationale. In this case, the pilot of the Merlin intends to make the approach at Craig despite 100 feet ceilings with ¼ mile visibility.

Identify actions that can successfully control the change. In the situation being discussed, the pilot looks at success as meeting several objectives:

1. Being on time for Thanksgiving dinner

2. Not inconveniencing his relatives waiting on the ground

3. Meeting his own predisposed objective of landing at Craig

The pilot failed to be objective in this case. The identification of courses of action were for his psychological success and not the safety of his family.

Do take the necessary action. In this case, the pilot contaminates his decision-making process and selects an approach to the instrument landing system (ILS) runway 32 at Craig where the weather was reported and observed far below the minimums.

Evaluate the effect of the action. In many cases like this, the pilot is so sure of his or her decision that the evaluation phase of his or her action is simply on track and on glideslope, despite impossible conditions. Because the situation seems in control, no other evaluation of the progress is employed.

The outcome of this accident was predictable considering the motivation of the pilot and his failure to monitor the approach using standard and accepted techniques. It was ascertained that the pilot, well above the decision height, saw a row of lights to his right that was interpreted as the runway environment. Instead of confirming with his aircraft's situational position, the pilot instead took over manually and flew toward the lights, descended below the glidepath, and impacted terrain. The passengers survived, but the pilot was killed.

Automatic Decision-Making

In an emergency situation, a pilot might not survive if he or she rigorously applied analytical models to every decision made; there is not enough time to go through all the options. But under these circumstances, does he or she find the best possible solution to every problem?

For the past several decades, research into how people actually make decisions has revealed that when pressed for time, experts faced with a task loaded with uncertainty, first assess whether the situation strikes them as familiar. Rather than comparing the pros and cons of different approaches, they quickly imagine how one or a few possible courses of action in such situations will play out. Experts take the first workable option they can find. While it may not be the best of all possible choices, it often yields remarkably good results.

The terms naturalistic and automatic decision-making have been coined to describe this type of decision-making. These processes were pioneered by Mr. Gary Kleinn, a research psychologist famous for his work in the field of automatic/naturalistic decision-making. He discovered that laboratory models of decision-making could not describe decision-making under uncertainty and fast dynamic conditions. His processes have influenced changes in the ways the Marines and Army train their officers to make decisions and are now impacting decision-making as used within the aviation environment. The ability to make automatic decisions holds true for a range of experts from fire fighters to police officers. It appears the expert's ability hinges on the recognition of patterns and consistencies that clarify options in complex situations. Experts appear to make provisional sense of a situation, without actually reaching a decision, by launching experience-based actions that in turn trigger creative revisions.

This is a reflexive type of decision-making anchored in training and experience and is most often used in times of emergencies when there is no time to practice analytical decision-making. Naturalistic or automatic decision-making improves with training and experience, and a pilot will find himself or herself using a combination of decision-making tools that correlate with individual experience and training. *Figure 5-2* illustrates the differences between traditional, analytical decision-making and naturalistic decision-making, both related to human behavior. Instances of human factor accidents include operational errors that relate to loss of situational awareness and flying outside the envelope. These can be termed as operational pitfalls.

Operational Pitfalls

Operational pitfalls are traps that pilots fall into, avoidance of which is actually simple in nature. A pilot should always have an alternate flight plan for where to land in case of an emergency on every flight. For example, a pilot may decide to spend a morning flying the traffic pattern but does not top off the fuel tanks because he or she is only flying the traffic pattern. Make considerations for the unexpected. What if another aircraft blows a tire during landing and the runway is closed? What will the pilot in the traffic pattern do? Although the odds may be low for something of this nature to happen, every pilot should have an alternate plan that answers the question, "Where can I land?" and the follow-up question, "Do I have enough fuel?"

Weather is the largest single cause of aviation fatalities. Most of these accidents occur to a GA operator, usually flying a light single- or twin-engine aircraft, who encounters

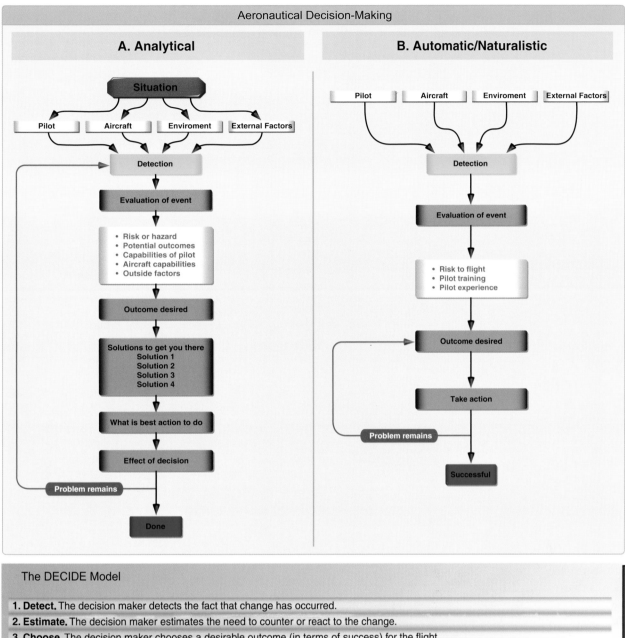

A. Analytical

Situation

Pilot | Aircraft | Enviroment | External Factors

Detection

Evaluation of event

- Risk or hazard
- Potential outcomes
- Capabilities of pilot
- Aircraft capabilities
- Outside factors

Outcome desired

Solutions to get you there
Solution 1
Solution 2
Solution 3
Solution 4

What is best action to do

Effect of decision

Problem remains

Done

B. Automatic/Naturalistic

Pilot | Aircraft | Enviroment | External Factors

Detection

Evaluation of event

- Risk to flight
- Pilot training
- Pilot experience

Outcome desired

Take action

Problem remains

Successful

The DECIDE Model

1. **Detect.** The decision maker detects the fact that change has occurred.
2. **Estimate.** The decision maker estimates the need to counter or react to the change.
3. **Choose.** The decision maker chooses a desirable outcome (in terms of success) for the flight.
4. **Identify.** The decision maker identifies actions which could successfully control the change.
5. **Do.** The decision maker takes the necessary action.
6. **Evaluate.** The decision maker evaluates the effect(s) of his/her action countering the change.

Figure 5-2. *The illustration on the left shows how the DECIDE model is used in decision-making and follows the five steps shown in the above left. In the automatic decision-making model (sometimes called naturalistic decision-making) the emphasis is recognizing a problem paired with a solution that is cultivated through both experience and training. In theory the automatic decision-making model seeks a quick decision at the cost of absolute accuracy where prolonged analysis is not practical. Naturalistic decision-making is generally used during emergencies where slow responsiveness is problematic and potentially additive to a problem.*

instrument meteorological conditions (IMC) conditions while operating under VFR. Over half the pilots involved in weather accidents did not receive an official weather briefing. Once the flight is under way, the number of pilots who receive a weather update from automated flight service station (AFSS) is dismal. An analysis done by FAA's Aviation Safety Information Analysis System (ASIAS) found that during a recent five-year period, only 19 pilots out of 586 fatal accident flights received any information from flight watch or an AFSS, once en route. It is important to recognize weather presents a hazard, which in turn can become an unmanageable risk. GA aircraft travel slowly and must fly in the weather rather than above it. Since weather is unpredictable, it is highly likely that during a flight, a pilot will encounter weather conditions different from what he or she expected. These weather conditions are not necessarily severe, like ice or thunderstorms, and analysis has shown that most VFR encounters with IMC involved low clouds and restrictions to visibility.

Scud Running

Scud running, or continued VFR flight into instrument flight rules (IFR) conditions, pushes the pilot and aircraft capabilities to the limit when the pilot tries to make visual contact with the terrain. This is one of the most dangerous things a pilot can do and illustrates how poor ADM links directly to a human factor that leads to an accident. A number of instrument-rated pilots die scud running while operating VFR. Scud running is seldom successful, as can be seen in the following accident report.

A Cessna 172C, piloted by a commercial pilot, was substantially damaged when it struck several trees during a precautionary landing on a road. Instrument meteorological conditions (IMC) prevailed at the time of the accident. The personal cross-country flight was being conducted without a flight plan.

The pilot had purchased the airplane in Arkansas and was ferrying it to his fixed base operation (FBO) in Utah. En route stops were made and prior to departing the last stop, the pilot, in a hurry and not wanting to walk back to the FBO to call flight service, discussed the weather with a friend who told the pilot that the weather was clear to the north. Poor weather conditions prevented him from landing at his original destination, so the pilot turned around and landed at a privately owned airport with no service facilities. Shortly thereafter, the pilot took off again and looped north toward his destination. The "weather got bad" and the pilot decided to make a precautionary landing on a snow-covered road. The road came to a "T" and the airplane slid off the end. The left wing and propeller struck the ground and the right wing struck a tree. The right wing had leading edge compression

damage outboard of the root, and the left wing leading edge was crushed near the wing tip fairing. Both propeller blades were bent. As discussed throughout this handbook, this accident was the result of a chain of poor decisions. The pilot himself recalled what he should have done in this situation, "I should have picked a spot to do a precautionary landing sooner before the weather got bad. Second, I should have called flight service to get a weather briefing, instead of discussing it with a friend on the ramp."

Get-There-Itis

In get-there-itis, personal or external pressure clouds the vision and impairs judgment by causing a fixation on the original goal or destination combined with a total disregard for alternative course of action.

"I have to be in Houston by 7 o'clock." In the previous case, the pilot was simply lazy.

Approximately 15 minutes after departure, the pilot of a Piper PA-34-200T twin-engine airplane encountered IMC. The non-instrument-rated private pilot lost control of the airplane and impacted snow-covered terrain. Prior to the cross-country flight, the pilot obtained three standard weather briefings, of which two were obtained on the previous day and one on the morning of the accident. The briefings included IFR conditions along the planned route of flight.

According to the briefing and a statement from a friend, the pilot intended to land the airplane prior to his destination if the weather conditions were not visual flight rules (VFR). The pilot would then "wait it out" until the weather conditions improved. According to radar data, the airplane departed from the airport and was traveling on a southeasterly heading. For the first 15 minutes of the flight, the airplane maintained a level altitude and a consistent heading. For the last minute of the flight, the airplane entered a descent of 2,500 feet per minute (fpm), a climb of 3,000 fpm, a 1,300 fpm descent, and the airplane's heading varied in several degrees. The airplane impacted the terrain in a right wing low, nose-down attitude.

Looking beyond the summary, get-there-itis leads to a poor aeronautical decision because this pilot repeatedly sought weather briefings for a VFR flight from Pueblo, Colorado, to Tyler, Texas. During a 17-minute briefing at 0452, he was informed of weather conditions along his planned route of flight that included IFR conditions that were moving south, moderate icing conditions for the state of Colorado, and low ceilings of visibility along the planned route of flight. His next call took place at 0505, approximately 1½ hours prior to takeoff. The pilot responded to the reported weather conditions by saying "so I've got a, I've got a little tunnel there that looks decent right now...from what that will tell me I've got a, I've got an open shot over the butte."

The pilot began the flight 1½ hours after his weather update, neglecting to weigh the risks created by a very volatile weather situation developing across the state.

The National Transportation Safety Board (NTSB) determined the probable cause of this accident was the pilot's failure to maintain control of the airplane after an inadvertent encounter with IMC, resulting in the subsequent impact with terrain. Contributing factors were the pilot's inadequate preflight planning, self-induced pressure to conduct the flight, and poor judgment.

Unfortunately for this pilot, he fell into a high-risk category. According to the NTSB, pilots on flights of more than 300 nautical miles (NM) are 4.7 times more likely to be involved in an accident than pilots on flights of 50 NM or less. Another statistic also put him in to the potential accident category: his lack of an instrument rating. Studies have found that VFR pilots are trained to avoid bad weather and when they find themselves in poor weather conditions, they do not have the experience to navigate their way through it.

Continuing VFR into IMC

Continuing VFR into IMC often leads to spatial disorientation or collision with ground/obstacles. It is even more dangerous when the pilot is not instrument rated or current. The FAA and NTSB have studied the problem extensively with the goal of reducing this type of accident. Weather-related accidents, particularly those associated with VFR flight into IMC, continue to be a threat to GA safety because 80 percent of the VFR-IMC accidents resulted in a fatality.

One question frequently asked is whether or not pilots associated with VFR flight into IMC even knew they were about to encounter hazardous weather. It is difficult to know from accident records exactly what weather information the pilot obtained before and during flight, but the pilot in the following accident departed in marginal visual meteorological conditions (VMC).

In 2007, a Beech 836 TC Bonanza was destroyed when it impacted terrain. The private, non-instrument-rated pilot departed in VMC on a personal flight and requested VFR flight following to his destination. When he neared his destination, he contacted approach control and reported that his altitude was 2,500 feet above mean sea level (MSL). Approach control informed the pilot that there were moderate to heavy rain showers over the destination airport. The pilot reported that he was experiencing "poor visibility" and was considering turning 180° to "go back." Approach control informed the

pilot that IMC prevailed north of his position with moderate to heavy rain showers. Their exchange follows:

At 1413:45, approach control asked the pilot if he was going to reverse course. The pilot replied, "Ah, affirmative, yeah we're gonna make, we're gonna actually head, ah, due north."

Approach control instructed the pilot to proceed to the northeast and maintain VFR.

At 1414:53, approach control asked the pilot what was his current destination. The pilot responded, "We're deviating. I think we're going to go back over near Eau Claire, but, ah, we're going to see what the weather is like. We're, we're kinda in the soup at this point so I'm trying to get back, ah, to the east."
At 1415:10, approach control informed the pilot that there was "some level one rain or some light rain showers" that were about seven miles ahead of his present position.

At 1415:30, the pilot asked approach control, "What is the ah, ah, Lakeville weather? I was showing seven thousand and overcast on the system here. Is that still holding?"

Approach control responded, "No, around [the] Minneapolis area we're overcast at twenty three hundred and twenty one hundred in the vicinity of all the other airports around here."

At 1415:49, the pilot stated, "I'm going to head due south at this time, down to, ah, about two thousand and make it into Lakeville."

Approach control responded, "...you can proceed south bound."

At 1416:02, the pilot responded, "...thanks (unintelligible)."

The radar data indicated that the airplane's altitude was about 2,600 feet MSL.

There were no further radio transmissions. After the last radio transmission, three radar returns indicated the airplane descended from 2,500 feet to 2,300 feet MSL before it was lost from radar contact.

A witness reported he heard an airplane and then saw the airplane descending through a cloud layer that was about 400–500 feet above the ground. The airplane was in about a 50° nose-down attitude with its engine producing "cruise power." He reported the airplane was flying at a high rate of speed for about four seconds until he heard the airplane impact the terrain. The observed weather in the area of the accident was reported as marginal VMC and IMC.

The NTSB determined the probable cause(s) of this accident to be the pilot's continued flight into IMC, which resulted in spatial disorientation and loss of control.

Research can offer no single explanation to account for this type of accident. Is it the end result of poor situational awareness, hazardous risk perception, motivational factors, or simply improper decision-making? Or is it that adequate weather information is unavailable, simply not used, or perhaps not understood? Extracting critical facts from multiple sources of weather information can be challenging for even the experienced aviator. And once the pilot is in the air, en route weather information is available only to the extent that he or she seeks it out if their aircraft is not equipped with operational weather displays.

No one has yet determined why a pilot would fly into IMC when limited by training to fly under VFR. In many cases, the pilot does not understand the risk. Without education, we have a fuzzy perception of hazards. It should be noted that pilots are taught to be confident when flying. Did overconfidence and ability conflict with good decision-making in this accident? Did this pilot, who had about 461 flight hours, but only 17 hours in make and model overrate his ability to fly this particular aircraft? Did he underestimate the risk of flying in marginal VFR conditions?

Loss of Situational Awareness

Situational awareness is the accurate perception and understanding of all the factors and conditions within the four fundamental risk elements (pilot, aircraft, environment, and type of operation) that affect safety before, during, and after the flight. Thus, loss of situational awareness results in a pilot not knowing where he or she is, an inability to recognize deteriorating circumstances, and the misjudgment of the rate of deterioration.

In 2007, an instrument-rated commercial pilot departed on a cross-country flight through IMC. The pilot made radio transmissions to ground control, tower, low radar approach control, and high radar approach control that he was "new at instruments" and that he had not flown in IMC "in a long time." While maneuvering to get back on the centerline of the airway, while operating in an area of heavy precipitation, the pilot lost control of the airplane after he became spatially disoriented.

Recorded radar data revealed flight with stable parameters until approximately 1140:49 when the airplane is recorded making an unexpected right turn at a rate of 2° per second.

The pilot may not have noticed a turn at this rate since there were no radar calls to departure control. The right turn continues until radar contact is lost at 1141:58 at which point that airplane is turning at a rate of approximately 5° per second and descending at over 3,600 fpm.

Wreckage and impact information was consistent with a right bank, low-angle, high-speed descent. IMC prevailed in the area at the time of the accident. The descent profile was found to be consistent with the "graveyard spiral." Prior to flight, for unknown reasons, the telephone conversations with the AFSS progressed from being conservative to a strong desire to fly home, consistent with the pilot phenomena "get-home-itis."

The 26 year old pilot was reported to have accumulated a total of 456.7 hours, of which 35.8 hours were in the same make and model. Prior to the accident flight, the pilot had accumulated a total of 2.5 hours of actual instrument time, with 105.7 hours of simulated instrument time.

The following abbreviated excerpt from the accident report offers insight into another example of poor aeronautical decision-making.

The pilot had telephoned AFSS six times prior to take off to request weather reports and forecasts. The first phone call lasted approximately 18 minutes during which time the AFSS briefer forecasts IMC conditions for the route of flight and briefs an airmen's meteorological information (AIRMET) for IFR conditions. The pilot stated that he did not try to take off a day earlier because he recalled that his instrument flight instructor told him not to take off if he did not feel comfortable.

During the second phone call, the pilot stated he was instrument rated but did not want to take any chances. At this time, the AFSS briefer forecast light rain and marginal conditions for VFR. The third phone call lasted approximately 5 minutes during which the AFSS briefer gives weather, the AIRMET, and forecasts a cycle of storms for the day of flight. The pilot responds that it sounds like a pretty bad day to fly. During the fourth phone call, the pilot states that he has been advised by a flight instructor at his destination airport that he should try to wait it out because the weather is "pretty bad right now" The AFSS briefer agrees and briefs light to moderate rain showers in the destination area and the AIRMET for IFR conditions. The AFSS briefer states that after 1100 the weather should improve.

At 1032, the pilot calls AFSS again and sounds distressed. The pilot stated he wants to get home, has not showered in 1½ days, is getting tired, and wants to depart as soon as possible.

The AFSS briefer briefs the AIRMET for IFR conditions and forecasts IFR en route. At 1055, the pilot phones AFSS for the final time and talks for approximately 7 minutes. Then, he files an IFR flight plan. The AFSS briefer states improving conditions and recommends delaying departure to allow conditions to improve. However, this pilot made the decision to fly in weather conditions clearly outside his personal flying comfort zone. Once he had exceeded his proficiency level, the newly minted instrument pilot had no instructor in the other seat to take over.

The NTSB determines the probable cause of this accident to be pilot loss of control due to spatial disorientation. Contributing factors were the pilot's perceived need to fly to home station and his lack of flight experience in actual IMC.

Flying Outside the Envelope

Flying outside the envelope is an unjustified reliance on the mistaken belief that the airplane's high performance capability meets the demands imposed by the pilot's (usually overestimated) flying skills. While it can occur in any type aircraft, advanced avionics aircraft have contributed to an increase in this type accident.

According to the Aircraft Owners and Pilots Association (AOPA) Air Safety Foundation (ASF), advanced avionics aircraft are entering GA fleet in large numbers. Advanced avionics aircraft includes a variety of aircraft from the newly designed to retrofitted existing aircraft of varying ages. What they all have in common are complex avionics packages. While advanced avionics aircraft offer numerous safety and operational advantages, the FAA has identified a safety issue that concerns pilots who develop an unwarranted overreliance on the avionics and the aircraft, believing the equipment will compensate fully for pilot shortcomings.

Related to overreliance is the role of ADM, which is probably the most significant factor in the GA accident record of high-performance aircraft used for cross-country flight. The FAA advanced avionics aircraft safety study found that poor decision-making seems to afflict new advanced avionics aircraft pilots at a rate higher than for GA as a whole. This is probably due to increased technical capabilities, which tempt pilots to operate outside of their personal (or even legal) limits. The availability of global positioning system (GPS) and moving map systems, coupled with traffic and near real-time weather information in the flight deck, may lead pilots to believe they are protected from the dangers inherent to operating in marginal weather conditions.

While advanced flight deck technologies may mitigate certain risks, it is by no means a substitute for sound ADM. The challenge is this: How should a pilot use this new information in flight to improve the safety of flight operations? The answer to this question lies in how well the pilot understands the information, its limitations, and how best to integrate this data into the ADM process.

According to AOPA, government information gathering on accidents does not contain definitive ways to differentiate between advanced avionics aircraft and non-advanced avionics aircraft; however, it is known that the aircraft in the following accident was an advanced avionics aircraft.

In 2003, during a cross-country flight, the non-instrument-rated private pilot encountered heavy fog and poor visibility. The airplane was destroyed after impacting the terrain in a wildlife refuge. Wildlife refuge personnel stated the weather was clear on the morning of the accident. However, later that morning, the weather deteriorated, and the wildlife refuge personnel stated, "the fog was very heavy and visibility was very poor."

An AIRMET, issued and valid for the area, reported the following: "occasional ceiling below 1,000 feet, visibility below 3 miles in mist, fog ... Mountains occasionally obscured clouds, mist, fog ..." On the day of the accident, the pilot did not file a flight plan or receive a formal weather briefing from an AFSS.

Examining this accident in more detail offers insight into the chain of events that led to this accident.

1. On the morning of the flight, the pilot used the Internet to complete three sessions with the Direct User Access Terminal Service (DUATS), filing his VFR flight plan during the third session. He departed in VFR conditions and requested and received VFR flight following until he approached a mountain range at which point he canceled his flight following services and continued en route without further FAA contact.

2. During the last leg of his flight, the pilot initiated a right turn of about 120°. This turn, which he initiated about 3,600 feet MSL, resulted in the aircraft flying along a narrow valley toward up-sloping terrain. The pilot continued in that direction for another 2 minutes before colliding with a number of trees near the top of a ridge.

The NTSB determines the probable cause(s) of this accident as follows: The pilot's inadvertent flight into IMC and failure to maintain clearance with the terrain. A contributing factor was the pilot's failure to obtain an updated preflight weather briefing.

The ASF offered the following comment for educational purposes: the non-instrument-rated pilot in this accident may or may not have been tempted to continue his flight when encountering IMC conditions because he had advanced avionics aircraft equipment on board.

3P Model

Making a risk assessment is important, but in order to make any assessment the pilot must be able to see and sense surroundings and process what is seen before performing a corrective action. An excellent process to use in this scenario is called the 3 Ps: Perceive, Process, and Perform.

The Perceive, Process, Perform (3P) model for ADM offers a simple, practical, and systematic approach that can be used during all phases of flight. *[Figure 5-3]* To use it, the pilot will:

- Perceive the given set of circumstances for a flight.
- Process by evaluating their impact on flight safety.
- Perform by implementing the best course of action.

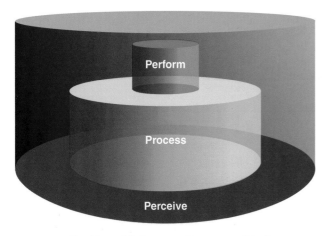

Figure 5-3. *The 3P model: Perceive, Process, and Perform.*

Examine a pilot flying into a canyon. Many pilots fail to see the difference between a valley and a canyon. Most valleys can be characterized as depressions with a predominant direction. A canyon is also a valley, but it is a very deep valley bordered by cliffs. One can infer that making a turn across a valley will be over rising terrain whose slope is shallow. A canyon, however, is bordered by vertical walls. Additionally, valleys are typically wider than canyons. However, before proceeding it is important to understand the relationship between rate of turn and turn radius.

Rate of Turn

The rate of turn (ROT) is the number of degrees (expressed in degrees per second) of heading change that an aircraft makes. The ROT can be determined by taking the constant of 1,091, multiplying it by the tangent of any bank angle and dividing that product by a given airspeed in knots as illustrated in *Figure 5-4*. If the airspeed is increased and the ROT desired is to be constant, the angle of bank must be increased; otherwise, the ROT decreases. Likewise, if the airspeed is held constant, an aircraft's ROT increases if the bank angle is increased. The formula in *Figures 5-4* through *5-6* depicts the relationship between bank angle and airspeed as they affect the ROT.

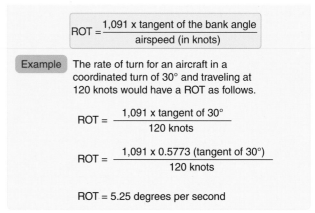

Figure 5-4. *Rate of turn for a given airspeed (knots, TAS) and bank angle.*

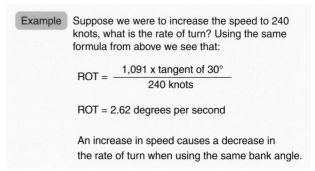

Figure 5-5. *Rate of turn when increasing speed.*

NOTE: All airspeeds discussed in this section are true airspeed (TAS).

Airspeed significantly affects an aircraft's ROT. If airspeed is increased, the ROT is reduced if using the same angle of bank used at the lower speed. Therefore, if airspeed is increased as illustrated in *Figure 5-5,* it can be inferred that the angle of bank must be increased in order to achieve the same ROT achieved in *Figure 5-6.*

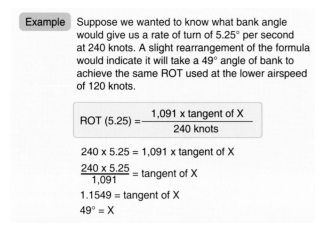

Suppose we wanted to know what bank angle would give us a rate of turn of 5.25° per second at 240 knots. A slight rearrangement of the formula would indicate it will take a 49° angle of bank to achieve the same ROT used at the lower airspeed of 120 knots.

$$\text{ROT (5.25)} = \frac{1{,}091 \times \text{tangent of X}}{240 \text{ knots}}$$

$240 \times 5.25 = 1{,}091 \times \text{tangent of X}$

$\dfrac{240 \times 5.25}{1{,}091} = \text{tangent of X}$

$1.1549 = \text{tangent of X}$

$49° = X$

Figure 5-6. *To achieve the same rate of turn of an aircraft traveling at 120 knots, an increase of bank angle is required.*

What does this mean on a practicable side? If a given airspeed and bank angle produces a specific ROT, additional conclusions can be made. Knowing the ROT is a given number of degrees of change per second, the number of seconds it takes to travel 360° (a circle) can be determined by simple division. For example, if moving at 120 knots with a 30° bank angle, the ROT is 5.25° per second and it takes 68.6 seconds (360° divided by 5.25 = 68.6 seconds) to make a complete circle. Likewise, if flying at 240 knots TAS and using a 30° angle of bank, the ROT is only about 2.63° per second and it takes about 137 seconds to complete a 360° circle. Looking at the formula, any increase in airspeed is directly proportional to the time the aircraft takes to travel an arc.

So, why is this important to understand? Once the ROT is understood, a pilot can determine the distance required to make that particular turn, which is explained in radius of turn.

Radius of Turn

The radius of turn is directly linked to the ROT, which is a function of both bank angle and airspeed, as explained earlier. If the bank angle is held constant and the airspeed is increased, the radius of the turn changes (increases). A higher airspeed causes the aircraft to travel through a longer arc due to a greater speed. An aircraft traveling at 120 knots is able to turn a 360° circle in a tighter radius than an aircraft traveling at 240 knots. In order to compensate for the increase in airspeed, the bank angle would need to be increased.

The radius of turn (ROT) can be computed using a simple formula. The radius of turn is equal to the velocity squared (V^2) divided by 11.26 times the tangent of the bank angle.

$$R = \frac{V^2}{11.26 \times \text{tangent of the bank angle}}$$

Using the examples provided in *Figures 5-4* through *5-6*, both the radii of the two speeds postulated can be computed. Noteworthy, is if the speed is doubled, the radius is squared. *[Figures 5-7 and 5-8]*

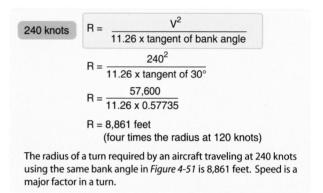

120 knots

$$R = \frac{V^2}{11.26 \times \text{tangent of bank angle}}$$

$$R = \frac{120^2}{11.26 \times \text{tangent of 30°}}$$

$$R = \frac{14{,}400}{11.26 \times 0.5773}$$

$$R = 2{,}215 \text{ feet}$$

The radius of a turn required by an aircraft traveling at 120 knots and using a bank angle of 30° is 2,215 feet.

Figure 5-7. *Radius at 120 knots.*

240 knots

$$R = \frac{V^2}{11.26 \times \text{tangent of bank angle}}$$

$$R = \frac{240^2}{11.26 \times \text{tangent of 30°}}$$

$$R = \frac{57{,}600}{11.26 \times 0.57735}$$

$$R = 8{,}861 \text{ feet}$$
(four times the radius at 120 knots)

The radius of a turn required by an aircraft traveling at 240 knots using the same bank angle in *Figure 4-51* is 8,861 feet. Speed is a major factor in a turn.

Figure 5-8. *Radius at 240 knots.*

In *Figure 5-9,* two aircraft enter a canyon. One aircraft enters at 120 knots, and the other at 140 knots. Both pilots realize they are in a blind canyon and need to conduct a course reversal. Both pilots perceive their unique environment and sense that something is occurring. From this perception, the pilots process the information, and then act. Although one may sense that this is similar to the DECIDE model, it is not. The 3P process is a continuous loop of the pilot's handling of hazards. The DECIDE model and naturalistic decision-making focus on particular problems requiring resolution. Therefore, pilots exercise the 3P process continuously, while the DECIDE model and naturalistic decision-making result from the 3P process.

Perceive

In the first step, the goal is to develop situational awareness by perceiving hazards, which are present events, objects, or circumstances that could contribute to an undesired future event. Both pilots realize they need to turn 180° for continued safe flight. The pilot systematically identifies and lists hazards associated with all aspects of the situation, and must do it fast and accurately.

Figure 5-9. *Two aircraft have flown into a canyon by error. The canyon is 5,000 feet across and has sheer cliffs on both sides. The pilot in the top image is flying at 120 knots. After realizing the error, the pilot banks hard and uses a 30° bank angle to reverse course. This aircraft requires about 4,000 feet to turn 180°, and makes it out of the canyon safely. The pilot in the bottom image is flying at 140 knots and also uses a 30° angle of bank in an attempt to reverse course. The aircraft, although flying just 20 knots faster than the aircraft in the top image, requires over 6,000 feet to reverse course to safety. Unfortunately, the canyon is only 5,000 feet across and the aircraft will hit the canyon wall. The point is that airspeed is the most influential factor in determining how much distance is required to turn. Many pilots have made the error of increasing the steepness of their bank angle when a simple reduction of speed would have been more appropriate.*

Process

In the second step, the goal is to process learned and practiced information to determine whether the identified hazards constitute risk, which is defined as the future impact of a hazard that is not controlled or eliminated. The degree of risk posed by a given hazard can be measured in terms of exposure or potential mishap and death.

The pilot flying at 120 knots is familiar with the formulas discussed before or is aware that slower speeds result in a smaller turning radius. The pilot flying at 140 knots does not slow down as he thinks that a 30° bank is satisfactory.

Perform

In both cases, the pilots perform the turns. The pilot performing a turn at 120 knots exits the canyon safely; while the pilot flying at 140 knots hits the canyon wall, killing all onboard. Another area, although not a canyon, is flying around buildings. Just a few years ago, a pilot collided with a building during a turn. Had he slowed down, he would be alive today.

The 3P model is intended to be a constant loop within which the pilot measures his or her actions through perception of the current, dynamically changing situation. Failure to do so results in error, an accident, and possible death. The pilot flying at 140 knots failed in this endeavor and paid the ultimate price. Therefore, the 3P process must be a continuous loop providing anomalies or reassurance that what is going on is what was predicted or unexpected.

Chapter Summary

The study of ADM, its history, and models for decision-making while in flight is only a precursor to its practical application. Regurgitating the meaning of the concepts allows a pilot to pass a checkride and written examination, but understanding is what saves lives and improves flight skills. Therefore, one can say that understanding these concepts is superior to being able to state them in a precise order or with absolute accuracy.

Chapter 6
Single-Pilot Resource Management

Introduction

While crew resource management (CRM) focuses on pilots operating in crew environments, many of the concepts apply to single pilot operations. Many CRM principles have been successfully applied to single-pilot aircraft and led to the development of single-pilot resource management (SRM). SRM is defined as the art of managing all the resources (both onboard the aircraft and from outside sources) available to a pilot prior to and during flight to ensure a successful flight. SRM includes the concepts of aeronautical decision-making (ADM), risk management, controlled flight into terrain (CFIT) awareness, and situational awareness. SRM training helps the pilot maintain situational awareness by managing automation, associated aircraft control, and navigation tasks. This enables the pilot to accurately assess hazards, manage resulting risk potential, and make good decisions.

Advanced System

SRM helps pilots learn to execute methods of gathering information, analyzing it, and making decisions. Although the flight is coordinated by a single person and not an onboard flightcrew, the use of available resources, such as air traffic control (ATC) and automated flight service stations (AFSS), replicates the principles of CRM.

Recognition of Hazards

As will be seen in the following accident, it is often difficult for the pilot involved to recognize a hazard and understand the risk. How a pilot interprets hazards is an important component of risk assessment. Failure to recognize a hazard becomes a fatal mistake in the following accident involving an experimental airplane.

During a cross-country night flight, an experimental airplane experienced an inflight fire followed by a loss of control. The aircraft hit a building and both the commercial pilot and the private pilot-rated passenger were killed. There were no injuries to anyone on the ground. Night visual meteorological conditions prevailed at the time. The flight departed from its home airport about 20:00. The experimental four-place, four-door, high-wing airplane had a composite fuselage powered by a Lycoming IO-360 engine. The aircraft had logged 94.1 hours.

At the time, the flight was transitioning through Class B airspace and receiving visual flight rules (VFR) advisories from Approach Control. According to the facility transcript, at 20:33:36 the pilot queried the controller about a fire smell and asked if there were fire activity in the marshland below them. The controller indicated in the negative, to which the pilot responded, "We just want know if it's the airplane that smells or the air." *[Figure 6-1]*

Shortly afterward, the pilot was advised of a frequency change, which was acknowledged. At 20:36:06, the pilot checked in with another controller and was given the current altimeter setting. A little more than 1½ minutes later, the controller transmitted that he was not receiving the airplane's Mode C transponder altitude, to which there was no response from the pilot. All communications with the aircraft were lost.

Radar data indicated that when the pilot queried the controller about a fire, the airplane was at 5,500 feet mean sea level (MSL) heading north. The airplane's radar track continued northbound until 20:37:13, at which time the last transponder return from the airplane was recorded. The remainder of the radar track (primary targets only) showed the airplane turning right to a heading of east-southeast. At about 20:39:20, the airplane turned further right to a heading of south. The last

Figure 6-1. *The pilot perceived something was wrong (see Chapter 5, Aeronautical Decision-Making) but failed to process the information correctly.*

Figure 6-2. *The pilot must consider all aspects of the flight to include form, fit, and function.*

radar return was received at 20:39:36. Three minutes later, the controllers were notified by police that an airplane had crashed into a building.

One witness reported that the airplane was flying at an altitude of about 500 feet above ground level (AGL) in a southeast direction when it made "a slight right turn, then a slight left turn, then a sharp right turn, then descended in what appeared to be in excess of 30° nose down." A second witness observed the airplane at an altitude of less than 100 feet AGL "in an excessive nose-down attitude towards the ground." Both witnesses reported that a large post-impact fire erupted.

The pilot, seated in the right front seat, held a commercial pilot certificate with airplane single- and multi-engine land and instrument ratings. Additionally, he held a flight instructor certificate with airplane single-engine land and instrument airplane ratings. According to Federal Aviation Administration (FAA) records, the pilot had accumulated a total flight time of over 1,400 hours. The passenger, who was seated in the left front seat, held a private pilot certificate with an airplane single-engine land rating. Records indicated the passenger was 6 feet 3 inches tall and weighed 231 pounds. *[Figure 6-2]*

The airplane was constructed by its manufacturer as a prototype for an experimental amateur-built kit and was issued a special airworthiness certificate in the category of experimental research and development. Material examination of the engine and propeller indicated no pre-accident discrepancies, and all major structures were accounted for. It was not possible to assess control continuity due to impact and subsequent fire.

Upon interview, representatives of the manufacturer indicated that the original pilot (left) seat in the airplane was replaced by the owner about a month prior to the accident with a six-way power seat from an automobile. *[Figure 6-3]* It was installed to accommodate customer requests for an adjustable seat. This seat incorporated three motors that facilitated the six-way movement of the seat. In its original automotive installation, it was wired using a 30-amp circuit breaker for protection; if any motor failed, the automobile circuit would trip. As installed in the automobile, if the breaker did not trip, the switch itself would fail. The seat was installed in the airplane with a 5-amp circuit breaker, but shortly after

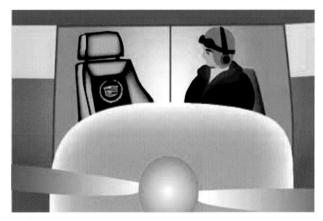

Figure 6-3. *This pilot simply wanted to be comfortable while flying.*

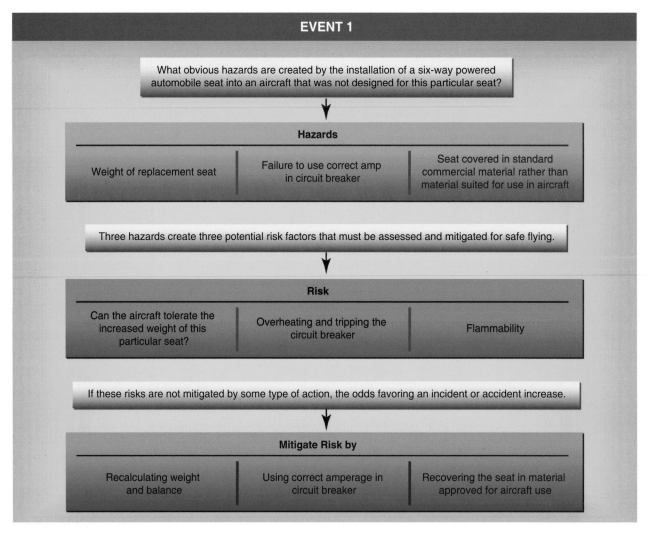

EVENT 1

What obvious hazards are created by the installation of a six-way powered automobile seat into an aircraft that was not designed for this particular seat?

Hazards

| Weight of replacement seat | Failure to use correct amp in circuit breaker | Seat covered in standard commercial material rather than material suited for use in aircraft |

Three hazards create three potential risk factors that must be assessed and mitigated for safe flying.

Risk

| Can the aircraft tolerate the increased weight of this particular seat? | Overheating and tripping the circuit breaker | Flammability |

If these risks are not mitigated by some type of action, the odds favoring an incident or accident increase.

Mitigate Risk by

| Recalculating weight and balance | Using correct amperage in circuit breaker | Recovering the seat in material approved for aircraft use |

Figure 6-4. *Example of an event diagram mapping hazards, risk assessment, and risk mitigation for the first event.*

installation, it was noted that a larger person in the left seat would trip the circuit breaker and the motors became hot. The 5-amp circuit breaker was replaced with a 7-amp circuit breaker to prevent excessive tripping.

The event diagram in *Figures 6-4* and *6-5* maps the hazards, risk assessment, and attempts to mitigate this accident.

As this accident demonstrates, for the pilot of an experimental aircraft, assessing risk goes beyond the self-assessment illustrated in the IMSAFE method. Hazard identification, risk assessment, and its mitigation starts much earlier. The construction method of manufacture and the materials used impose a certain inherent risk that may not be apparent until an adverse event occurs. Unfortunately, hindsight is of limited value to the aircraft passengers and pilot, but do provide others a better understanding of risk and its insidious nature.

The risk assessment matrix in *Figure 6-6* can provide lessons from this accident. The vertical scale relates to the likelihood

of something happening, while the horizontal scale indicates impact upon safety of the flight.

While impact damage precluded the National Transportation Safety Board (NTSB) from determining the cause of the fire for the aircraft involved in this accident, the final report discusses the possibility that one of the motors to the seat overheated and ignited the seat cushion. They attributed this possibility to the circuit breaker issue as well as the past instance of the circuit breaker tripping when a large occupant sat in the seat.

It is probable that the installation of the replacement seat started a chain of events diagramed above that led to a fatal accident. The three hazards associated with the seat are discussed more fully below:

1. Effect of weight on the aircraft weight and balance and its downstream performance—a seat with three motors adds significant weight on one side to the aircraft.

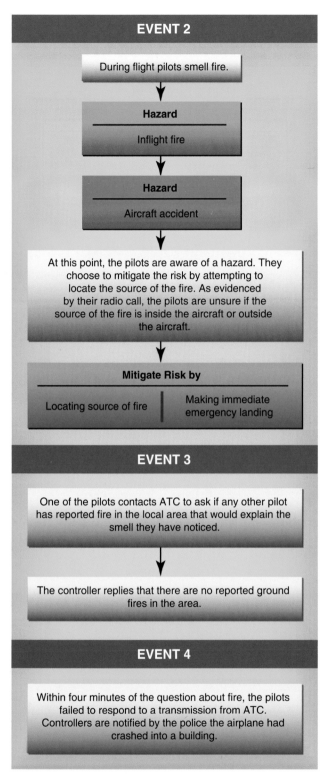

EVENT 2

During flight pilots smell fire.

Hazard

Inflight fire

Hazard

Aircraft accident

At this point, the pilots are aware of a hazard. They choose to mitigate the risk by attempting to locate the source of the fire. As evidenced by their radio call, the pilots are unsure if the source of the fire is inside the aircraft or outside the aircraft.

Mitigate Risk by

| Locating source of fire | Making immediate emergency landing |

EVENT 3

One of the pilots contacts ATC to ask if any other pilot has reported fire in the local area that would explain the smell they have noticed.

The controller replies that there are no reported ground fires in the area.

EVENT 4

Within four minutes of the question about fire, the pilots failed to respond to a transmission from ATC. Controllers are notified by the police the airplane had crashed into a building.

Figure 6-5. *Event diagram for events 2 through 4.*

Even with weight allowances, aircraft performance would be affected.

2. Seat materials—the criteria for automobile materials are different from those for materials suitable for use

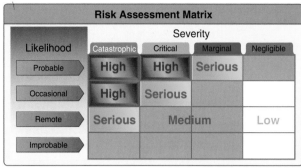

Figure 6-6. *The installation of non-aviation parts can have a profound effect.*

in aircraft. Material coverings certified for aircraft use provide additional safety and are intended to reduce unnecessary exposure to fire. In this accident, the possibility exists that the seat covering on the automotive seat exacerbated the fire.

3. Potential for electrical malfunctions, especially overheating—why use a 5-amp and then a 7-amp circuit breaker when a 30-amp circuit breaker was used in the original automotive installation?

Did the pilot in command (PIC) take unnecessary risk? Assuming he or she had no knowledge of the differences between the replacement seat and a normal aircraft seat, he should have questioned the installation of a non-aircraft part. And, examine the PIC's query to the controller during the flight. He indicated he was not sure if his aircraft were on fire or if something on the ground were on fire. Did he incorrectly assess the information he had been given? Did he assume his aircraft was not on fire? Given the seat's installation, its propensity to overheat, and the indication of a fire, what should the pilot have done?

In *Figure 6-6,* the risk matrix relates directly to both the builder of the aircraft and the PIC.

*   Builder—the likelihood of an adverse event is minimized when aviation standards are adopted in both the selection of material and components, and their installation. The more closely the standards are followed, the less likely the occurrence of an adverse event. In this case, the likelihood of an adverse event is maximized not only because of the seat installation, but that it represents a potential problem across the construction of the entire aircraft.

*   PIC—if he were familiar with the seat installation, the problems it created, and its prior problem of overheating, he failed to assess the likelihood that the source of the smell was a fire in the aircraft and not a fire on the ground. No information is available

on how long the occupants of the aircraft smelled the smoke, but there were only four minutes between the radio call requesting information about ground fires and the impact with the building. This left the pilot little time to react to a hazard that metamorphosed into a catastrophe.

Rating the likelihood of an impending problem means a pilot needs to ask key questions. For instance, the PIC of this accident needed to ask the aircraft builder how the addition of this seat affected the aircraft. "If this component fails, what are the consequences or severity of the problems it creates?" Obviously, the installation of this seat produced issues in many areas: the seat cover material, electrical loading, weight and balance, and the impact of the added weight upon aircraft performance. Independently, these factors may not create an catastrophic hazard, but taken collectively, they can create a chain of failures that lead to a fatal accident.

The PIC recognized a fire was in evidence while in flight. Given aviation historical data regarding inflight fires, smoke in the flight deck is considered an emergency. In this case, the controller even eliminated one source as a possibility. He told the pilot no ground fires had been reported. Did the PIC fail to take seriously that the smoke must be from his aircraft? Did this pilot make a poor inflight decision or did he make a poor preflight decision?

This example illustrates how an aircraft that is not constructed to standards places the unaware pilot with an element of risk. In 1983, an amateur builder in Alabama used improper wing bolts to secure his homebuilt's wings. The manufacturer called for the use of eight special close-tolerance high-strength bolts that cost approximately 40 dollars each. The homebuilder found what he decided were the same bolts at his local farm supply center for less than 2 dollars each. Upon takeoff, the bolts sheared at about 15 feet in altitude. Consequently, the aircraft's wings collapsed, causing permanent disability to the pilot as a result of his injuries. The bolts he used were simple, low-strength material bolts used for wooden gates.

Use of Resources

To make informed decisions during flight operations, a pilot must also become aware of the resources found inside and outside the flight deck. Since useful tools and sources of information may not always be readily apparent, learning to recognize these resources is an essential part of ADM training. Resources must not only be identified, but a pilot must also develop the skills to evaluate whether there is time to use a particular resource and the impact its use has upon the safety of flight. For example, the assistance of ATC may be very useful if a pilot becomes lost, but in an emergency situation, there may be no time to contact ATC.

During an emergency, a pilot makes an automatic decision and prioritizes accordingly. Calling ATC may take away from time available to solve the problem. Ironically, the pilot who feels the hourglass is running out of sand would be surprised at the actual amount of time available in which to make decisions. The perception of "time flying" or "dragging" is based upon various factors. If the pilot were to repeat the event (in which time seemed to evaporate) but had been briefed on the impending situation and could plan for it, the pilot would not feel the pressure of time "flying." This example demonstrates the theory that proper training and physiological well-being is critical to pilot safety.

Internal Resources

One of the most underutilized resources may be the person in the right seat, even if the passenger has no flying experience. When appropriate, the PIC can ask passengers to assist with certain tasks, such as watching for traffic or reading checklist items. *[Figure 6-7]*

A passenger can assist the PIC by:

- Providing information in an irregular situation, especially if familiar with flying. A strange smell or sound may alert a passenger to a potential problem.

- Confirming after the pilot that the landing gear is down.

- Learning to look at the altimeter for a given altitude in a descent.

- Listening to logic or lack of logic.

Also, the process of a verbal briefing (which can happen whether or not passengers are aboard) can help the PIC in the decision-making process. For example, assume a pilot provides his passenger a briefing of the forecasted landing weather before departure. When the Automatic Terminal Information Service (ATIS) is picked up at the destination and the weather has significantly changed, the integration of this report and forecasted weather causes the pilot to explain to a passenger the significance or insignificance of the disparity. The pilot must provide a cohesive analysis and explanation that is understood by the passenger. Telling passengers everything is okay when the weather is ¼ mile away is not fooling anyone. Therefore, the integration of briefing passengers is of great value in giving them a better understanding of a situation. Other valuable internal resources include ingenuity, solid aviation knowledge, and flying skill.

When flying alone, another internal resource is verbal communication. It has been established that verbal communication reinforces an activity; touching an object while communicating further enhances the probability an

Figure 6-7. *When possible, have a passenger reconfirm that critical tasks are completed.*

activity has been accomplished. For this reason, many solo pilots read the checklist out loud; when they reach critical items, they touch the switch or control. For example, to ascertain the landing gear is down, the pilot can read the checklist and hold the gear handle down until there are three green lights. This tactile process of verbally communicating coupled with a physical action are most beneficial.

It is necessary for a pilot to have a thorough understanding of all the equipment and systems in the aircraft being flown. Lack of knowledge, such as knowing if the oil pressure gauge is direct reading or uses a sensor, is the difference between making a wise decision or poor one that leads to a tragic error.

Checklists are essential flight deck internal resources. They are used to verify that aircraft instruments and systems are checked, set, and operating properly. They also ensure the proper procedures are performed if there is a system malfunction or inflight emergency. Students reluctant to use checklists can be reminded that pilots at all levels of experience refer to checklists, and that the more advanced the aircraft is, the more crucial checklists become. In addition, the pilot's operating handbook (POH) is required to be carried on board the aircraft and is essential for accurate flight planning and resolving inflight equipment malfunctions. However, the ability to manage workload is the most valuable resource a pilot has. *[Figure 6-8]*

Figure 6-8. *The pilot must continually juggle various facets of flight, which can become overwhelming. The ability to prioritize, manage inflight challenges, and digest information makes the pilot a better professional.*

External Resources

Air traffic controllers and AFSS are the best external resources during flight. In order to promote the safe, orderly flow of air traffic around airports and along flight routes, the ATC provides pilots with traffic advisories, radar vectors, and assistance in emergency situations. Although it is the PIC's responsibility to make the flight as safe as possible, a pilot with a problem can request assistance from ATC. [Figure 6-9] For example, if a pilot needs to level off, be given a vector, or decrease speed, ATC assists and becomes integrated as part of the crew. The services provided by ATC can not only decrease pilot workload, but also help pilots make informed inflight decisions.

Figure 6-9. *Controllers work to make flights as safe as possible.*

The AFSS are air traffic facilities that provide pilot briefing, en route communications, VFR search and rescue services, assist lost aircraft and aircraft in emergency situations, relay ATC clearances, originate Notices to Airmen (NOTAM), broadcast aviation weather and National Airspace System (NAS) information, receive and process IFR flight plans, and monitor navigational aids (NAVAIDs). In addition, at selected locations, AFSS provide En Route Flight Advisory Service (Flight Watch), issue airport advisories, and advise Customs and Immigration of transborder flights. Selected AFSS in Alaska also provide Transcribed Weather En Route Broadcast (TWEB) recordings and take weather observations.

Another external resource available to pilots is the very high frequency (VHF) Direction Finder (VHF/DF). This is one of the common systems that helps pilots without their awareness of its operation. FAA facilities that provide VHF/DF service are identified in the airport/facility directory (A/FD). DF equipment has long been used to locate lost aircraft and to guide aircraft to areas of good weather or to airports. DF instrument approaches may be given to aircraft in a distress or urgent condition.

Experience has shown that most emergencies requiring DF assistance involve pilots with little flight experience. With this in mind, DF approach procedures provide maximum flight stability in the approach by using small turns and wings-level descents. The DF specialist gives the pilot headings to fly and tells the pilot when to begin a descent. If followed, the headings lead the aircraft to a predetermined point such as the DF station or an airport. To become familiar with the procedures and other benefits of DF, pilots are urged to request practice DF guidance and approaches in VFR weather conditions.

SRM and the 5P Check

SRM is about how to gather information, analyze it, and make decisions. Learning how to identify problems, analyze the information, and make informed and timely decisions is not as straightforward as the training involved in learning specific maneuvers. Learning how to judge a situation and "how to think" in the endless variety of situations encountered while flying out in the "real world" is more difficult.

There is no one right answer in ADM, rather each pilot is expected to analyze each situation in light of experience level, personal minimums, and current physical and mental readiness level, and make his or her own decision.

SRM sounds good on paper, but it requires a way for pilots to understand and use it in their daily flights. One practical application is called the Five Ps (5 Ps). [Figure 6-10] The 5 Ps are:

- Plan
- Plane
- Pilot
- Passengers
- Programming

Each of these areas consists of a set of challenges and opportunities that face a single pilot. Each can substantially increase or decrease the risk of successfully completing the flight based on the pilot's ability to make informed and timely decisions. The 5 Ps are used to evaluate the pilot's current situation at key decision points during the flight or when an emergency arises. These decision points include preflight, pretakeoff, hourly or at the midpoint of the flight, predescent, and just prior to the final approach fix or for VFR operations, just prior to entering the traffic pattern.

Figure 6-10. *The 5 Ps.*

The first decision is whether to go or not to go on the flight, and the easiest point at which to cancel due to bad weather is the evening before the scheduled flight. A good pilot always watches the weather and checks weather information sources to stay abreast of current conditions and forecasts. This enables him or her to warn passengers that the weather conditions are questionable and they might need a backup plan. The subsequent visit to the flight planning room (or call to AFSS) provides all the information readily available to make a sound decision, and is where communication and Fixed Base Operator (FBO) services are readily available to make alternate travel plans. *[Figures 6-11 and 6-12]*

For instance, the easiest point to cancel a flight due to bad weather is before the pilot and passengers walk out the door and load the aircraft. So, the first decision point is preflight in the flight planning room.

The 5 Ps are based on the idea that the pilots have essentially five variables that impact their environment and can cause the pilot to make a single critical decision or several less critical decisions that when added together can create a critical outcome. This concept stems from the belief that current decision-making models tended to be reactionary in nature. A change has to occur and be detected to drive a risk management decision by the pilot. For instance, many pilots use risk management sheets that are filled out by the pilot prior to takeoff. These form a catalog of risks that may be encountered that day and turn them into numerical values. If the total exceeds a certain level, the flight is altered or cancelled. Informal research shows that while these are useful documents for teaching risk factors, they are almost never used outside of formal training programs. The 5P concept is an attempt to take the information contained in those sheets and in other available models and put it to good use.

Figure 6-12. *The first decision point is during the preflight planning.*

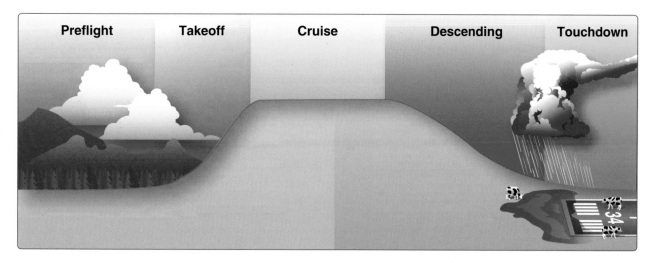

Figure 6-11. *The 5Ps are applied to various modes prior to and during the flight.*

The second easiest point in the flight to make a critical safety decision is just prior to takeoff. Few pilots have ever had to make an emergency takeoff. While the point of the 5P check is to help the pilot fly, the correct application of the 5 Ps before takeoff is to assist in making a reasoned go/no-go decision based on all the information available. The decision is usually to go with certain restrictions and changes but may also be a no-go. The key fact is that these two points in the process of flying are critical go/no-go points on each and every flight. *[Figure 6-13]*

Figure 6-13. *This is a good time to assess self and fatigue. Are you running late? Have you checked the destination weather yet? Remember that when things are going wrong, they do not get better with time.*

The third point at which to review the 5 Ps is the midpoint of the flight. *[Figure 6-14]* Pilots often wait until the ATIS is in range to check weather, yet at this point in the flight many good options have already been passed. Additionally, fatigue and low altitude hypoxia serve to rob the pilot of

Figure 6-14. *The second point in applying the 5P check is just before takeoff.*

much of his or her energy by the end of a long and tiring flight day. Fatigue affects memory, attention to detail, and communication ability. Frequently associated with pilot error, it also impairs coordination and degrades situational awareness, seriously influencing a pilot's ability to make effective decisions. There are several types fatigue. Physical fatigue results from sleep loss, exercise, or physical work while factors such as stress and prolonged performance of cognitive work result in mental fatigue.

Hypoxia or oxygen starvation also robs a pilot of physical and mental acuity. Oxygen deprivation is insidious because it sneaks up on the unwary and steals the first line of sensory protection, the sense that something is wrong. The human body does not give reliable signals at the onset of hypoxia so a pilot needs special training in how to recognize the symptoms. This training is important because the brain is the first part of the body to reflect a diminished oxygen supply and evidence of that is usually a loss of judgment.

Everyone's response to hypoxia varies, but the effects of hypoxia can be safely experienced under professional supervision at the Civil Aeromedical Institute's altitude chamber in Oklahoma City and at 14 cooperating military installations throughout the United States. To attend a 1-day physiological training course, contact the FAA Accident Prevention Specialist for an Aeronautical Center (AC) Form 3150-7.

Once a pilot begins to suffer a loss of energy, he or she transitions from a decision-making mode to an acceptance mode. If the flight is longer than 2 hours, the 5P check should be conducted hourly. This is also a good time to evaluate the destination airport. Believe it or not many pilots have more problems on the ground taxiing than on the approach. Because larger airports have taxiways designed for large transport aircraft, the vantage point for a 767 crew sitting 18 feet off the ground regarding taxiways (especially at night) is superior to that for a pilot of a Cessna 172 with a vantage point at 6 feet. Therefore, at the midpoint of the flight, the pilot should review the layout, approaches, and the taxiway structure and its identification system. For instance, at Atlanta Hartsfield, a pilot is expected to understand the difference between "inner and outer M" (Mike) taxiway, and at Dulles a pilot is expected to know where "spot two" is located. Landing is not the time to review the airport facility. Conversely, if a pilot does not know the idiosyncrasies of the airport, requesting progressive instructions and/or letting ATC know he or she is "not familiar" reflects professionalism.

The last two decision points are just prior to descent into the terminal area and just prior to the final approach fix, or if VFR just prior to entering the traffic pattern, as preparations for landing commence. Most pilots execute approaches with the expectation that they will land out of the approach every time. A healthier approach requires the pilot to assume that changing conditions (the 5 Ps) will cause the pilot to divert or execute the missed approach on every approach. This keeps the pilot alert to conditions that may increase risk and threaten the safe conduct of the flight. Diverting from cruise altitude saves fuel, allows unhurried use of the autopilot, and is less reactive in nature. Diverting from the final approach fix, while more difficult, still allows the pilot to plan and coordinate better rather than executing a futile missed approach. A detailed discussion of each of the 5 Ps follows.

Plan

The plan can also be called the mission or the task. It contains the basic elements of cross-country planning: weather, route, fuel, current publications, etc. The plan should be reviewed and updated several times during the course of the flight. *[Figure 6-15]* A delayed takeoff due to maintenance, fast-moving weather, and a short-notice temporary flight restriction (TFR) may all radically alter the plan. The plan is not only about the flight plan, but also all the events that surround the flight and allow the pilot to accomplish the mission. The plan is always being updated and modified and is especially responsive to changes in the other four remaining Ps. If for no other reason, the 5P check reminds the pilot that the day's flight plan is real life and subject to change at any time.

Obviously, weather is a huge part of any plan. The addition of real time data link weather information provided by advanced avionics gives the pilot a real advantage in inclement weather, but only if the pilot is trained to retrieve and evaluate the weather in real time without sacrificing situational awareness. And of course, weather information should drive a decision, even if that decision is to continue on the current plan. Pilots of aircraft without datalink weather should get updated weather in flight through an AFSS and/or Flight Watch.

Plane

Both the plan and the plane are fairly familiar to most pilots. The plane consists of the usual array of mechanical and cosmetic issues that every aircraft pilot, owner, or operator can identify. *[Figure 6-16]* With the advent of advanced avionics, the plane has expanded to include database currency, automation status, and emergency backup systems that were unknown a few years ago. Much has been written about single-pilot IFR flight both with and without an autopilot. While use of autopilot is a personal decision, it is just that—a decision. Low IFR in a non-autopilot equipped aircraft may depend on several of the other Ps to be discussed. Pilot proficiency, currency, and fatigue are among them.

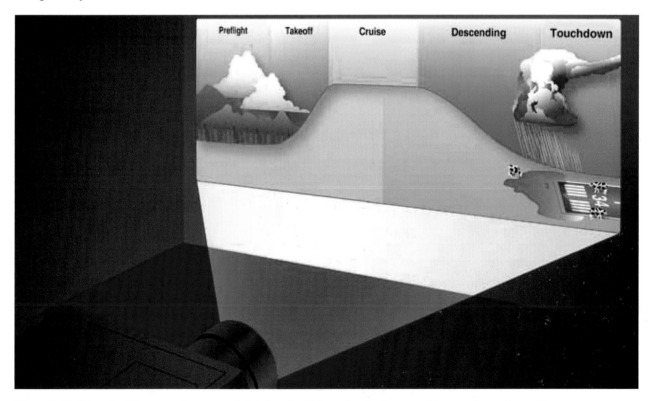

Figure 6-15. *Using the 5P process does not end with the takeoff. It needs to be integrated into routine workmanship.*

Figure 6-16. *The plane consists of not only the normal mechanical components but also the many advanced systems and software that supports it.*

Pilot

Flying, especially when used for business transportation, can expose the pilot to high altitude flying, long distance and endurance, and more challenging weather. An advanced avionics aircraft, simply due to its advanced capabilities, can expose a pilot to even more of these stresses. The traditional "IMSAFE" checklist is a good start. *[Figure 6-17]*

The combination of late night, pilot fatigue, and the effects of sustained flight above 5,000 feet may cause pilots to become less discerning, less critical of information, less decisive, and more compliant and accepting. Just as the most critical portion of the flight approaches (e.g., a night instrument approach in weather after a 4-hour flight), the pilot's guard is down the most. The 5P process helps a pilot recognize the physiological situation at the end of the flight before takeoff and continues to update personal conditions as the flight progresses. Once risks are identified, the pilot is better equipped to make alternate plans that lessen the effects of these factors and provide a safer solution.

Passengers

One of the key differences between CRM and SRM is the way passengers interact with the pilot. The pilot of a single-

✓ **I'M SAFE CHECKLIST**

Illness—Do I have any symptoms?

Medication—Have I been taking prescription or over-the-counter drugs?

Stress—Am I under psychological pressure from the job? Worried about financial matters, health problems, or family discord?

Alcohol—Have I been drinking within 8 hours? Within 24 hours?

Fatigue—Am I tired and not adequately rested?

Emotion—Am I emotionally upset?

Figure 6-17. *Making sure a pilot is ready to perform to a high standard is as important as the aircraft—maybe more!*

engine aircraft enters into a very personal relationship with the passengers. In fact, the pilot and passengers sit within arm's reach all of the time. *[Figure 6-18]*

Figure 6-18. *Passengers can be used effectively within the flight deck; simple things such as keeping an eye out for other aircraft is invaluable.*

If the capabilities of a passenger sitting next to the pilot are not being utilized, the pilot is limiting the potential for a successful flight. Passengers can read checklists, verify PIC performance of an action, re-verify that the gear is down and the lights are on, look for other aircraft, and even tune radios. The failure of a pilot to integrate the passenger at some level of assistance is almost as bad as not utilizing a pilot in that seat. Another person onboard is a resource for the PIC to use. A bonus is heightened passenger appreciation for GA through the participation in the flight.

Sometimes passengers also have their own priorities that influence the PIC. The desire of the passengers to make airline connections or important business meetings easily enters into a pilot's decision-making loop. Done in a healthy and open way, this can be a positive factor. Consider a flight to Dulles Airport and the passengers, both close friends and business partners, need to get to Washington, D.C., for an important meeting. The weather is VFR all the way to southern Virginia then turns to low IFR as the pilot approaches Dulles. A pilot employing the 5P approach might consider reserving a rental car at an airport in northern North Carolina or southern Virginia to coincide with a refueling stop. Thus, the passengers have a way to get to Washington, and the pilot has an alternate plan to avoid being pressured into continuing the flight if the conditions do not improve.

Passengers can also be pilots. If no one is designated as pilot in command (PIC) and unplanned circumstances arise, the decision-making styles of several self-confident pilots may come into conflict. Pilots also need to understand that non-pilots may not understand the level of risk involved in the flight. There is an element of risk in every flight. That is why SRM calls it risk management, not risk elimination. While a pilot may feel comfortable with the risk present in a night IFR flight, the passengers may not. A pilot employing SRM should ensure the passengers are involved in the decision-making and given tasks and duties to keep them busy and involved. If, upon a factual description of the risks present, the passengers decide to buy an airline ticket or rent a car, then a good decision has generally been made. This discussion also allows the pilot to move past what he or she thinks the passengers want to do and find out what they actually want to do. This removes self-induced pressure from the pilot.

Programming

The advanced avionics aircraft adds an entirely new dimension to the way GA aircraft are flown. The electronic instrument displays, GPS, and autopilot reduce pilot workload and increase pilot situational awareness. *[Figure 16-19]* While programming and operation of these devices are fairly simple and straightforward unlike the analog instruments they replace, they tend to capture the pilot's attention and hold it for long periods of time. To avoid this phenomenon, the pilot should plan in advance when and where the programming for approaches, route changes, and airport information gathering should be accomplished, as well as times it should not. Pilot familiarity with the equipment, the route, the local air traffic control environment, and personal capabilities vis-à-vis the automation should drive when, where, and how the automation is programmed and used.

Figure 6-19. *Understanding automation requires not just familiarization with the concepts but thorough understanding of the different systems.*

The pilot should also consider what his or her capabilities are in response to last minute changes of the approach (and the reprogramming required) and ability to make large-scale changes (a reroute for instance) while hand flying the aircraft. Since formats are not standardized, simply moving from one manufacturer's equipment to another should give the pilot pause and require more conservative planning and decisions.

Chapter Summary

The SRM process is simple. At least five times before and during the flight, the pilot should review and consider the plan, plane, pilot, passengers, and programming and make the appropriate decision required by the current situation. It is often said that failure to make a decision is a decision. Under SRM and the 5 Ps, even the decision to make no changes to the current plan is made through a careful consideration of all the risk factors present.

Chapter 7

Automation

Introduction

In the general aviation (GA) community, an automated aircraft is generally comprised of an integrated advanced avionics system consisting of a primary flight display (PFD), a multifunction display (MFD) including an instrument-certified global positioning system (GPS) with traffic and terrain graphics, and a fully integrated autopilot. This type of aircraft is commonly known as an advanced avionics aircraft. In an advanced avionics aircraft, the PFD is displayed on the left computer screen and the MFD is on the right screen.

Automation is the single most important advance in aviation technologies. Electronic flight displays (EFDs) have made vast improvements in how information is displayed and what information is available to the pilot. Pilots can access onboard information electronically that includes databases containing approach information, primary instrument display, and moving maps that mirror sectional charts, or display modes that provide three-dimensional views of upcoming terrain. These detailed displays depict airspace, including temporary flight restrictions (TFRs). MFDs are so descriptive that many pilots fall into the trap of relying solely on the moving maps for navigation. *[Figure 7-1]*

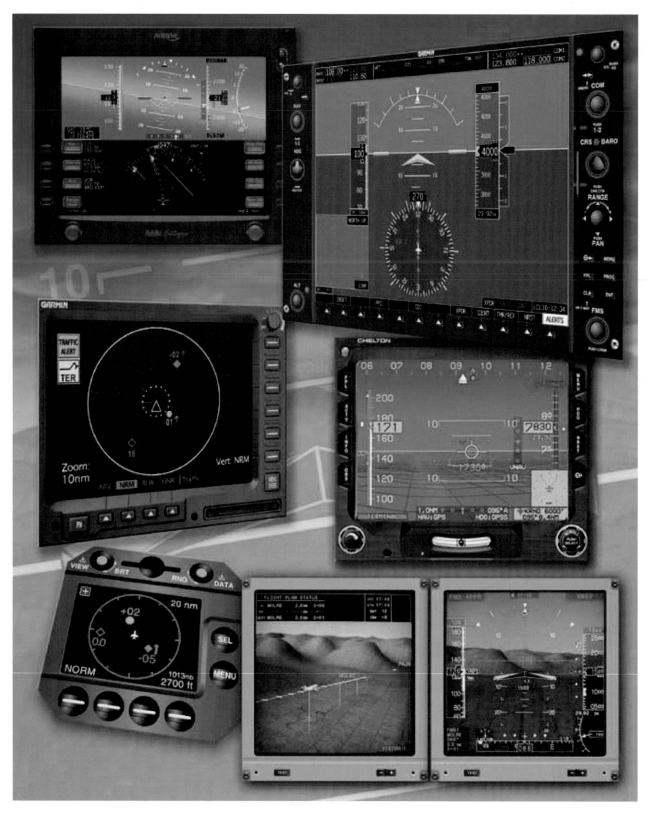

Figure 7-1. *Electronic flight instrumentation comes in many systems and provides a myriad of information to the pilot.*

More pilots now rely on automated flight planning tools and electronic databases for flight planning rather than planning the flight by the traditional methods of laying out charts, drawing the course, identifying navigation points (assuming a visual flight rules (VFR) flight), and using the pilot's operating handbook (POH) to figure out the weight and balance and performance charts. Whichever method a pilot chooses to plan a flight, it is important to remember to check and confirm calculations.

Most of the aviation community believes automation has made flying safer, but there is a fear that pilots fail to see that automation is a double-edged sword. Pilots need to understand the advantages of automation while being aware of its limitations. Experience has shown that automated systems can make some errors more evident while sometimes hiding other errors or making them less obvious. In 2005, the British Airline Pilots Association (BALPA) raised concerns about the way airline pilots are trained to depend upon automation. BALPA felt the current training leads to a lack of basic flying skills and inability to cope with an inflight emergency, especially mechanical failures. The union believes passenger safety could be at risk.

Cockpit Automation Study

Concerns about the effect of automation on flight skills are not new. In 1995, the erosion of manual flight skills due to automation was examined in a study designed by Patrick R. Veillette and R. Decker. Their conclusions are documented in "Differences in Aircrew Manual Skills and Automated and Conventional Flightdecks," published in the April 1995 edition of the Transportation Research Record, an academic journal of the National Research Council. In the February 2006 issue of Business and Commercial Aviation (BCA), Dr. Patrick R. Veillette returned to this topic in his article "Watching and Waning."

The Veillette-Decker seminal study on automation came at a time when automated flight decks were entering everyday line operations and concern was growing about some of the unanticipated side effects. Deterioration of basic pilot skills was one of these concerns. While automation made the promise of reducing human mistakes, in some instances it actually created larger errors. When this study was undertaken, the workload in an automated flight deck in the terminal environment actually seemed higher than in the older conventional flight decks. At other times, automation seemed to lull the flight crews into complacency. Fears arose that the manual flying skills of flight crews using automation deteriorated due to an overreliance on computers. In fact, BALPA voiced a fear that has dogged automation for years: that pilots using automation have less "stick and rudder" proficiency when those skills were needed to resume direct manual control of the aircraft.

Thus, the Veillette-Decker study sought to determine what, if any, possible differences exist in manual flight skills between aircrews assigned to conventional and automated flight decks. Limited to normal and abnormal operations in terminal airspace, it sought to determine the degree of difference in manual flying and navigational tracking skills. Commercial airline crew members flying the conventional transport aircraft or the automated version were observed during line-oriented flight training.

The data set included various aircraft parameters such as heading, altitude, airspeed, glideslope, and localizer deviations, as well as pilot control inputs. These were recorded during a variety of normal, abnormal, and emergency maneuvers during 4-hour simulator sessions. All experimental participants were commercial airline pilots holding airline transport pilot certificates. The control group was composed of pilots who flew an older version of a common twin-jet airliner equipped with analog instrumentation. The experimental group was composed of pilots who flew newer models of that same aircraft equipped with a first generation electronic flight instrument system (EFIS) and flight management system (FMS).

When pilots who had flown EFIS for several years were required to fly various maneuvers manually, the aircraft parameters and flight control inputs clearly showed some erosion of flying skills. During normal maneuvers, the EFIS group exhibited somewhat greater deviations than the conventional group. Most of the time, the deviations were within the Practical Test Standard (PTS), but the pilots definitely did not keep on the localizer and glideslope as smoothly as the conventional group. The differences in hand-flying skills between the two groups became more significant during abnormal maneuvers such as steeper than normal visual approaches (slam-dunks).

Analysis of the aircraft data consistently had pilots of automated aircraft exhibit greater deviations from assigned courses and aircraft state parameters, and greater deviations from normal pitch and bank attitudes, than the pilots of conventional flight deck aircraft. *[Figure 7-2]* The most significant differences were found to occur during the approach and landing phases. It is industry practice to tolerate very little air speed deviation from the recommended value during approach and landing. The FAA's Practical Test Standards (PTS) for the airline transport rating allow a final approach speed of no more than five knots faster than recommended.

Another situation used in the simulator experiment reflected real world changes in approach that are common and can be assigned on short notice. While a pilot's lack of familiarity with the EFIS is often an issue, the approach would have been made easier by disengaging the automated system and manually flying the approach.

The emergency maneuver, engine-inoperative instrument landing system (ILS) approach, continued to reflect the same performance differences in manual flying skills between the two groups. The conventional pilots tended to fly raw data and when given an engine failure, they performed it expertly. When EFIS crews had their flight directors disabled, their eye scan began a more erratic searching pattern and their manual flying subsequently suffered. According to Dr. Veillette's 2005 article, those who reviewed the data "saw that the EFIS pilots who better managed the automation also had better flying skills."

While the Veillette-Decker study offers valuable information on the effects of cockpit automation on the pilot and crew, experience now shows that increased workloads from advanced avionics results from the different timing of the manual flying workloads. Previously, the pilot(s) were busiest during takeoff and approach or landing. With the demands of automation programming, most of the workloads have been moved to prior to takeoff and prior to landing. Since Air Traffic Control (ATC) deems this the most appropriate time to notify the pilot(s) of a route or approach change, a flurry of reprogramming actions occurs at a time when management of the aircraft is most critical.

Reprogramming tasks during the approach to landing phase of flight can trigger aircraft mishandling errors that in turn snowball into a chain of errors leading to incidents or accidents. It does not require much time to retune a VOR for a new ILS, but it may require several programming steps to change the ILS selection in an FMS. In the meantime, someone must fly or monitor and someone else must respond to ATC instructions. In the pilot's spare time, checklists should be used and configuration changes accomplished and checked. Almost without exception, it can be stated that the faster a crew attempts to reprogram the unit, the more errors will be made.

Since publication of the Veillette-Decker study, increasing numbers of GA aircraft have been equipped with integrated advanced program avionics systems. These systems can lull pilots into a sense of complacency that is shattered by an inflight emergency. Thus, it is imperative for pilots to understand that automation does not replace basic flying skills. Automation adds to the overall quality of the flight experience, but it can also lead to catastrophe if not utilized properly. A moving map is not meant to substitute for a VFR sectional or low altitude en route chart. When using automation, it is recommended pilots use their best judgment and choose which level of automation will most efficiently do the task, considering the workload and situational awareness.

Pilots also need to maintain their flight skills and ability to maneuver aircraft manually within the standards set forth in the PTS. It is recommended that pilots of automated aircraft occasionally disengage the automation and manually fly the aircraft to maintain stick-and-rudder proficiency. In fact, a major airline recommends that their crews practice their instrument approaches in good weather conditions and use the autopilot in the bad weather conditions and monitor the flight's parameters.

More information on potential automation issues can be found at the flight deck automation issues website: www. flightdeckautomation.com. This website includes a searchable database containing over 1,000 records of data that support or refute 94 issues with automated flying.

Realities of Automation

Advanced avionics offer multiple levels of automation from strictly manual flight to highly automated flight. No one level of automation is appropriate for all flight situations, but in order to avoid potentially dangerous distractions when flying with advanced avionics, the pilot must know how to manage the course deviation indicator (CDI), navigation source, and the autopilot. It is important for a pilot to know the peculiarities of the particular automated system being used. This ensures the pilot knows what to expect, how to monitor for proper operation, and promptly take appropriate action if the system does not perform as expected.

For example, at the most basic level, managing the autopilot means knowing at all times which modes are engaged and which modes are armed to engage. The pilot needs to verify that armed functions (e.g., navigation tracking or altitude capture) engage at the appropriate time. Automation management is another good place to practice the callout

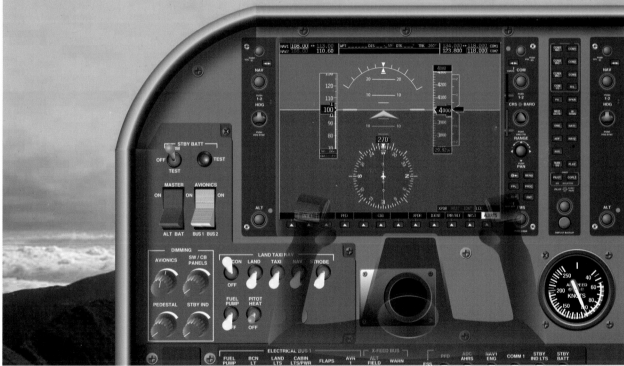

Figure 7-2. *Two flight decks equipped with the same information but in two different formats: analog and digital. What are they indicating? Chances are that the analog pilot will review the top display before the bottom display. Conversely, the digitally trained pilot will review the instrument panel on the bottom first.*

technique, especially after arming the system to make a change in course or altitude.

In advanced avionics aircraft, proper automation management also requires a thorough understanding of how the autopilot interacts with the other systems. For example, with some autopilots, changing the navigation source on the Electronic Horizontal Situation Indicator (e-HSI) from GPS to localizer (LOC) or VOR while the autopilot is engaged in NAV (course tracking mode) causes the autopilot's NAV mode to disengage. The autopilot's lateral control defaults to wings level until the pilot takes action to reengage the NAV mode to track the desired navigation source.

Enhanced Situational Awareness

An advanced avionics aircraft may offer increased safety with enhanced situational awareness. Although aircraft flight manuals (AFM) explicitly prohibit using the moving map, topography, terrain awareness, traffic, and weather datalink displays as the primary data source, these tools nonetheless give the pilot unprecedented information for enhanced situational awareness. Without a well-planned information management strategy, these tools also make it easy for an unwary pilot to slide into the complacent role of passenger in command.

Consider the pilot whose navigational information management strategy consists solely of following the magenta line on the moving map. He or she can easily fly into geographic or regulatory disaster if the straight line GPS course goes through high terrain or prohibited airspace or if the moving map display fails.

Risk is also increased when the pilot fails to monitor the systems. By failing to monitor the systems and failing to check the results of the processes, the pilot becomes detached from aircraft operation. This type of complacency led to tragedy in a 1999 aircraft accident in Colombia. A multi-engine aircraft crewed by two pilots struck the face of the Andes Mountains. Examination of their FMS revealed they entered a waypoint into the FMS incorrectly by one degree, resulting in a flightpath taking them to a point 60 nautical miles (NM) off the intended course. The pilots were equipped with the proper charts, their route was posted on the charts, and they had a paper navigation log indicating the direction of each leg. They had all the tools to manage and monitor their flight, but instead allowed the automation to fly and manage itself. The system did exactly what it was programmed to do; it flew on a programmed course into a mountain, resulting in multiple deaths. The pilots simply failed to manage the system and created their own hazard. Although this hazard was self-induced, what is notable is the risk the pilots created

through their own inattention. By failing to evaluate each turn made at the direction of automation, the pilots maximized risk instead of minimizing it. In this case, an avoidable accident became a tragedy through simple pilot error and complacency.

Not only did the crew fail to fully monitor the aircraft's automated routing, they also failed to retract the spoilers upon adding full thrust. This prevented the aircraft from outclimbing the slope of the mountain. Simulations of the accident indicate that had the aircraft had automatic spoiler retraction (spoilers automatically retract upon application of maximum thrust), or if the crew had remembered the spoilers, the aircraft probably would have missed the mountain.

Pilots en route to La Paz unwittingly deselected the very low frequency (VLF) input, thereby rendering the automation system unreliable. Although the system alerted the pilots to the ambiguity of navigation solution, the pilots perceived the alert to be computer error, and followed the course it provided anyway. They reached what they thought should be La Paz, but which was later estimated to have been approximately 30 NM away. They attempted to execute the published approach but were unable to tune the VOR radio, so they used instead the VLF of the KNS 660 to guide them on an impromptu approach. They were unable to gain visual contact with the runway environment due to in-cloud conditions despite the reported weather being clear with unrestricted visibility. Then they proceeded to their alternate about 1½ hours away. After 2½ hours of flight and following what they thought was the proper course, the aircraft became fuel critical, necessitating a controlled let-down from FL 250 to presumably visually conditions. Ironically, at about 9,000 mean sea level (MSL) they broke out of the cloud cover above an airfield. Although they attempted to align themselves for the runway, the aircraft ran out of fuel. The pilots dead-sticked the King Air to a ramp after which they broke through a fence, went over a berm, and into a pond. The aircraft was destroyed. After exiting the aircraft relatively unscathed, they found out they landed in Corumba, Brazil. [Figure 7-3]

In this accident, the pilots failed to realize that when no Omega signals were available with the VLF/Omega system, the equipment could continue to provide a navigation solution with no integrity using only the VLF system. Although the VLF/Omega system is now obsolete and has been replaced by the Global Navigation Satellite System (GNSS) and Loran-C, this accident illustrates the need for pilots of all experience levels to be thoroughly familiar with the operation of the avionics equipment being used. A pilot must not only know and understand what is being displayed, but must also be aware of what is not being displayed.

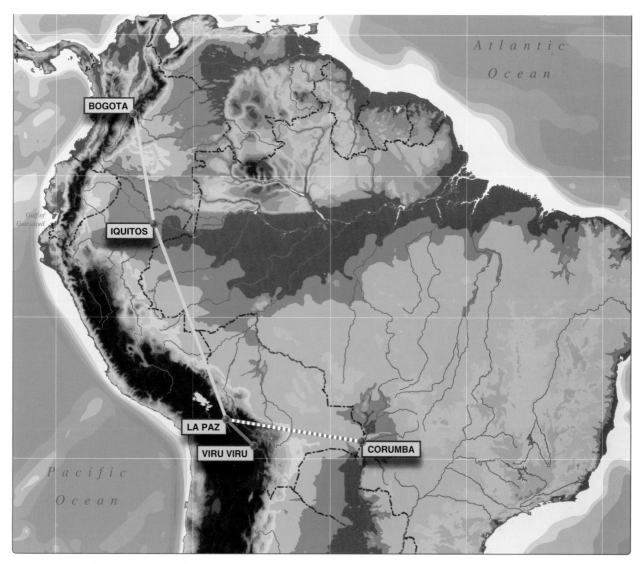

Figure 7-3. *The pilots of a King Air 200 had a flight from Bogota, Colombia, to Iquitos, Peru, (for fuel) and then to La Paz, Bolivia, as final destination. They listed Viru Viru (located at Santa Cruz, Bolivia) as their alternate. The aircraft was equipped with a Bendix King KNS 660 that provided integrated navigation solutions based upon VOR, DME, and two variants of VLF radios. At that time, GPS had not yet been integrated into the FMS.*

A good strategy for maintaining situational awareness of information management should include practices that help ensure that awareness is enhanced by the use of automation, not diminished. Two basic procedures are to always double-check the system and conduct verbal callouts. At a minimum, ensure the presentation makes sense. Was the correct destination fed into the navigation system? Callouts—even for single-pilot operations—are an excellent way to maintain situational awareness, as well as manage information.

Other ways to maintain situational awareness include:

- Performing a verification check of all programming. Before departure, check all information programmed while on the ground.

- Checking the flight routing. Before departure, ensure all routing matches the planned flight route. Enter the planned route and legs, to include headings and leg length, on a paper log. Use this log to evaluate what has been programmed. If the two do not match, do not assume the computer data is correct, double check the computer entry.

- Verifying waypoints.

- Making use of all onboard navigation equipment. For example, use VOR to back up GPS and vice versa.

- Matching the use of the automated system with pilot proficiency. Stay within personal limitations.

- Planning a realistic flight route to maintain situational awareness. For example, although the onboard equipment allows a direct flight from Denver, Colorado, to Destin, Florida, the likelihood of rerouting around Eglin Air Force Base's airspace is high.
- Being ready to verify computer data entries. For example, incorrect keystrokes could lead to loss of situational awareness because the pilot may not recognize errors made during a high workload period.

Autopilot Systems

In a single-pilot environment, an autopilot system can greatly reduce workload. *[Figure 7-4]* As a result, the pilot is free to focus attention on other flight deck duties. This can improve situational awareness and reduce the possibility of a controlled flight into terrain (CFIT) accident. While the addition of an autopilot may certainly be considered a risk control measure, the real challenge comes in determining the impact of an inoperative unit. If the autopilot is known to be inoperative prior to departure, this may factor into the evaluation of other risks.

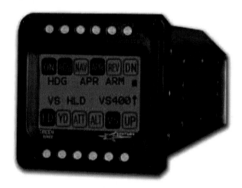

Figure 7-4. *An example of an autopilot system.*

For example, the pilot may be planning a VOR approach down to minimums on a dark night into an unfamiliar airport. In such a case, the pilot may have been relying heavily on a functioning autopilot capable of flying a coupled approach. This would free the pilot to monitor aircraft performance. A malfunctioning autopilot could be the single factor that takes this from a medium to a serious risk. At this point, an alternative needs to be considered. On the other hand, if the autopilot were to fail at a critical (high workload) portion of this same flight, the pilot must be prepared to take action. Instead of simply being an inconvenience, this could quickly turn into an emergency if not properly handled. The best way to ensure a pilot is prepared for such an event is to study the issue carefully prior to departure and determine well in advance how an autopilot failure is to be handled.

Familiarity

As previously discussed, pilot familiarity with all equipment is critical in optimizing both safety and efficiency. A pilot's being unfamiliar with any aircraft system will add to workload and may contribute to a loss of situational awareness. This level of proficiency is critical and should be looked upon as a requirement, not unlike carrying an adequate supply of fuel. As a result, pilots should not look upon unfamiliarity with the aircraft and its systems as a risk control measure, but instead as a hazard with high risk potential. Discipline is the key to success.

Respect for Onboard Systems

Automation can assist the pilot in many ways, but a thorough understanding of the system(s) in use is essential to gaining the benefits it can offer. Understanding leads to respect, which is achieved through discipline and the mastery of the onboard systems. However, it is important to fly the airplane without complete reliance on the PFD. This includes turns, climbs, descents, and flying approaches.

Reinforcement of Onboard Suites

The use of an electronic flight display (EFD) may not seem intuitive, but competency becomes better with understanding and practice. Computer-based software and incremental training help the pilot become comfortable with the onboard suites. Then, the pilot needs to practice what was learned in order to gain experience. Reinforcement not only yields dividends in the use of automation, it also reduces workload significantly.

Getting Beyond Rote Workmanship

The key to working effectively with automation is getting beyond the sequential process of executing an action. If a pilot has to analyze what key to push next, or always uses the same sequence of keystrokes when others are available, he or she may be trapped in a rote process. This mechanical process indicates a shallow understanding of the system. Again, the desire is to become competent and know what to do without having to think about "what keystroke is next." Operating the system with competency and comprehension benefits a pilot when situations become more diverse and tasks increase.

Understand the Platform

Contrary to popular belief, flight in aircraft equipped with different electronic management suites requires the same attention as aircraft equipped with analog instrumentation and a conventional suite of avionics. The pilot should review and understand the different ways in which EFDs are used in a particular aircraft. *[Figure 7-5]*

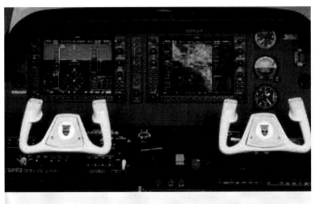

Figure 7-5. *Examples of different platforms. Top to bottom are the Beechcraft Baron G58, Cirrus SR22, and Cessna Entegra.*

Two simple rules for use of an EFD:

- Fly the aircraft to the standards in the PTS. Although this may seem insignificant, knowing how to fly the aircraft to a standard makes a pilot's airmanship smoother and allows for more time to attend to the system instead of managing multiple tasks.

- Read and understand the installed electronic flight system's manuals to include the use of the autopilot and the other onboard electronic management tools.

- Adhere to AFM/POH procedures.

Flight Management Skills

Automation Management

Before any pilot can master aircraft automation, he or she must first know how to fly the aircraft. Maneuver training remains an important component of flight training because almost 40 percent of all GA accidents take place in the landing phase, one realm of flight that still does not involve programming a computer to execute. Another 15 percent of all GA accidents occur during takeoff and initial climb.

An advanced avionics safety issue identified by the FAA concerns pilots who apparently develop an unwarranted overreliance in their avionics and the aircraft, believing that the equipment compensates for pilot shortcomings. Related to that overreliance is the role of ADM, which is probably the most significant factor in the GA accident record of high performance aircraft used for cross-country flight. The FAA advanced avionics aircraft safety study found that poor decision-making seems to afflict new advanced avionics pilots at a rate higher than that of GA as a whole. The review of advanced avionics accidents cited in this study shows the majority are not caused by something directly related to the aircraft, but by the pilot's lack of experience and a chain of poor decisions. One consistent theme in many of the fatal accidents is continued VFR flight into IMC.

Thus, pilot skills for normal and emergency operations hinge not only on mechanical manipulation of the stick and rudder, but also include the mental mastery of the EFD. Three key flight management skills are needed to fly the advanced avionics safely: information, automation, and risk.

Information Management

For the newly transitioning pilot, the PFD, MFD, and GPS/VHF navigator screens seem to offer too much information presented in colorful menus and submenus. In fact, the pilot may be drowning in information, but unable to find a specific piece of information. It might be helpful to remember these systems are similar to computers that store some folders on a desktop and some within a hierarchy.

The first critical information management skill for flying with advanced avionics is to understand the system at a conceptual level. Remembering how the system is organized helps the pilot manage the available information. It is important to understand that learning knob-and-dial procedures is not enough. Learning more about how advanced avionics systems work leads to better memory for procedures and allows pilots to solve problems they have not seen before.

There are also limits to understanding. It is impossible to understand all of the behaviors of a complex avionics system. Knowing to expect surprises and to continually learn new things is more effective than attempting to memorize mechanical manipulation of the knobs. Simulation software and books on the specific system used are of great value.

The second critical information management skill is to sense what is going on. Pilots new to advanced avionics often become fixated on the knobs and try to memorize each and every sequence of button pushes, pulls, and turns. A far better strategy for accessing and managing the information available in advanced avionics computers is to stop, look, and read. Reading before pushing, pulling, or twisting can often save a pilot some trouble.

Once in front of the display screens on an advanced avionics aircraft, the pilot must manage and prioritize the information flow to accomplish specific tasks. Certificated flight instructors (CFIs), as well as pilots transitioning to advanced avionics, will find it helpful to corral the information flow. This is possible through such tactics as configuring the aspects of the PFD and MFD screens according to personal preferences. For example, most systems offer map orientation options that include "north up," "track up," "desired track (DTK) up," and "heading up." Another tactic is to decide, when possible, how much (or how little) information to display. Pilots can also tailor the information displayed to suit the needs of a specific flight.

Information flow can also be managed for a specific operation. The pilot has the ability to prioritize information for a timely display of exact information needed for any given flight operation. Examples of managing information display for a specific operation include:

- Programming map scale settings for en route versus terminal area operation.

- Utilizing the terrain awareness page on the MFD for a night or IMC flight in or near the mountains.

- Using the nearest airports inset on the PFD at night or over inhospitable terrain.

- Programming the weather datalink set to show echoes and METAR status flags.

Risk Management

Risk management is the last of the three flight management skills needed for mastery of the advanced avionics aircraft. The enhanced situational awareness and automation capabilities offered by a glass flight deck vastly expand its safety and utility, especially for personal transportation use. At the same time, there is some risk that lighter workloads could lead to complacency.

Humans are characteristically poor monitors of automated systems. When passively monitoring an automated system for faults, abnormalities, or other infrequent events, humans perform poorly. The more reliable the system is, the worse the human performance becomes. For example, the pilot monitors only a backup alert system, rather than the situation that the alert system is designed to safeguard. It is a paradox of automation that technically advanced avionics can both increase and decrease pilot awareness.

It is important to remember that EFDs do not replace basic flight knowledge and skills. They are a tool for improving flight safety. Risk increases when the pilot believes the gadgets compensate for lack of skill and knowledge. It is especially important to recognize there are limits to what the electronic systems in any light GA aircraft can do. Being pilot in command (PIC) requires sound ADM, which sometimes means saying "no" to a flight.

For the GA pilot transitioning to automated systems, it is helpful to note that all human activity involving technical devices entails some element of risk. Knowledge, experience, and flight requirements tilt the odds in favor of safe and successful flights. The advanced avionics aircraft offers many new capabilities and simplifies the basic flying tasks, but only if the pilot is properly trained and all the equipment is working properly.

Pilot management of risk is improved with practice and consistent use of basic and practical risk management tools.

Chapter Summary

The advantages of automation are offset by its limitations. Accident data are used to explain enhanced situational awareness.

Chapter 8
Risk Management Training

Introduction

When introducing system safety to instructor pilots, the discussion invariably turns to the loss of traditional stick and rudder skills. The fear is that emphasis on items such as risk management, aeronautical decision-making (ADM), single-pilot resource management (SRM), and situational awareness detracts from the training that is so necessary in developing safe pilots. Also, because the Federal Aviation Administration's (FAA) current practical test standards (PTS) place so much emphasis on stick-and-rudder performance, there is concern that a shifting focus would leave flight students unprepared for that all-too-important check ride.

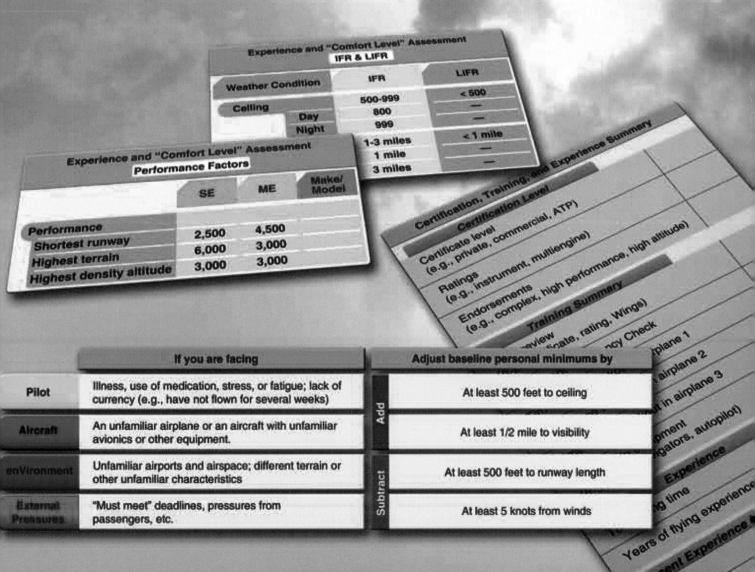

Experience and "Comfort Level" Assessment
IFR & LIFR

Weather Condition		IFR	LIFR
Ceiling		500-999	< 500
	Day	800	—
	Night	999	—
	1-3 miles		< 1 mile
	1 mile		—
	3 miles		—

Experience and "Comfort Level" Assessment
Performance Factors

	SE	ME	Make/Model
Performance	2,500	4,500	
Shortest runway	6,000	3,000	
Highest terrain	3,000	3,000	
Highest density altitude	3,000	3,000	

Certification, Training, and Experience Summary

Certification Level
Certificate level
(e.g., private, commercial, ATP)
Ratings
(e.g., instrument, multiengine)
Endorsements
(e.g., complex, high performance, high altitude)

Training Summary

	If you are facing		Adjust baseline personal minimums by
Pilot	Illness, use of medication, stress, or fatigue; lack of currency (e.g., have not flown for several weeks)	Add	At least 500 feet to ceiling
Aircraft	An unfamiliar airplane or an aircraft with unfamiliar avionics or other equipment.		At least 1/2 mile to visibility
enVironment	Unfamiliar airports and airspace; different terrain or other unfamiliar characteristics	Subtract	At least 500 feet to runway length
External Pressures	"Must meet" deadlines, pressures from passengers, etc.		At least 5 knots from winds

System Safety Flight Training

Instructors must understand that system safety flight training occurs in three phases. First, there are the traditional stick-and-rudder maneuvers. In order to apply the critical thinking skills that are to follow, pilots must first have a high degree of confidence in their ability to fly the aircraft. Next, the tenets of system safety are introduced into the training environment. In the manner outlined previously, students begin to learn how best to identify hazards, fully recognize all the risks with that hazard and manage or mitigate those risks, and use all available resources to make each flight as safe as possible. This can be accomplished through scenarios that emphasize the skill sets being taught. In the third phase, the student is introduced to more complex scenarios demanding focus on several safety-of-flight issues.

A traditional stick-and-rudder maneuver, such as a short-field landing, can be used to illustrate how ADM and risk management can be incorporated into instruction. In phase I, the initial focus is on developing the stick-and-rudder skills required to execute this operation safely. These include power and airspeed management, aircraft configuration, placement in the pattern, wind correction, determining the proper aim point and sight picture, etc. By emphasizing these points through repetition and practice, a student will eventually acquire the skills needed to execute a short-field landing.

Phase II introduces the many factors that come into play when performing a short-field landing, which include runway conditions, no-flap landings, airport obstructions, and rejected landings. The introduction of such items does not need to increase training times. In fact, all of the hazards or considerations referenced in the short-field landing lesson plan may be discussed in detail during the ground portion of the instructional program. For example, if training has been conducted at an airport which enjoys an obstruction free 6,000-foot runway, consider the implications of operating the same aircraft out of a 1,800 foot strip with an obstruction off the departure end. Add to that additional considerations, such as operating the aircraft at close to its maximum gross weight under conditions of high density altitude. Now, a single training scenario has several layers of complexity. The ensuing discussion proves a valuable training exercise, and it comes with little additional ground training and no added flight training.

Finally, phase III takes the previously discussed hazards, risks, and considerations and incorporates them into a complex scenario. This forces a student to consider not only a specific lesson item (in this case, short-field landings), but also requires that it be viewed in the greater context of the overall flight. For example, on a cross-country flight, the student is presented with a realistic distraction, perhaps the illness of a passenger. This forces a diversion to an alternate for which the student has not planned. The new destination airport has two runways, the longest of which is closed due to construction. The remaining runway is short, but while less-than-ideal, should prove suitable for landing. However, upon entering the pattern, the student finds the electrically driven flaps will not extend. The student must now consider whether to press on and attempt the landing or proceed to a secondary alternate.

If he or she decides to go forward and attempt the landing, this will prove an excellent time to test the requisite stick-and-rudder skills. If the student decides to proceed to a second alternate, this opens new training opportunities. Proceeding further tests cross-country skills, such as navigation, communication, management of a passenger in distress, as well as the other tasks associated with simply flying the aircraft. The outlined methodology simply takes a series of seemingly unrelated tasks and scripts them into a training exercise requiring both mechanical and cognitive skills for successful completion.

In addition, system safety may be applied to important safety lessons with less quantifiable performance standards. For example, controlled flight into terrain (CFIT) is an issue of concern to all pilots. In general aviation (GA), CFIT normally results from a combination of factors including weather, unfamiliar environment, nonstandard procedures, breakdown or loss of communications, loss of situational awareness, lack of perception of hazards, and lack of sound risk management techniques. Collectively, these conditions are difficult to replicate in most flight training environments. However, the subject may still be covered effectively during ground school and cross-country flight operations by using system safety methodology. Because CFIT is always the final "link" in the accident chain, it must be taught within the context of other flight operations; operations that increase the likelihood of a CFIT accident. This not only helps illustrate how easily these accidents can occur, it also highlights the conditions under which such accidents most often take place.

Other sources of risk management training available to the pilot are the various pilot organizations, such as the Airplane Owners and Pilots Association (AOPA), Experimental Aircraft Association (EAA), and numerous aircraft associations. All these organizations have variations of pilot experiences and solutions to situations in their publications. AOPA's Air Safety Foundation provides live seminars throughout the country and online training at the AOPA website: www. aopa.org.

Setting Personal Minimums

One of the most important concepts that safe pilots understand is the difference between what is "legal" in terms of the regulations, and what is "smart" or "safe" in terms of pilot experience and proficiency. By establishing personal minimums, pilots can take a big step in managing risk. In the article, "Getting the Maximum from Personal Minimums," (May/June 2006 FAA Aviation News), the FAA General Aviation and Commercial Division, AFS-800, discusses six steps for establishing personal minimums.

Step 1—Review Weather Minimums

Most people think of personal minimums primarily in terms of weather conditions, so begin with a quick review of weather definitions. The regulations define weather flight conditions for visual flight rules (VFR) and instrument flight rules (IFR) in terms of specific values for ceiling and visibility. *[Figure 8-1]*

IFR is defined as a ceiling less than 1,000 feet above ground level (AGL) and/or visibility less than three miles. Low instrument flight rules (LIFR) is a subcategory of IFR. VFR has ceiling greater than 3,000 feet AGL and visibility greater than five miles. Marginal visual flight rules (MVFR) is a subcategory of VFR.

Step 2—Assess Experience and Comfort Level

At first glance, this part of the process might look a bit complicated. It might take a few minutes to review, record, and summarize your personal experience, but you will find the finished product is well worth your time.

First, think back through your flight training and complete the Certification Training, an Experience Summary chart in *Figure 8-2*. The Certification, Training, and Experience Summary is adapted from the FAA's Personal and Weather Risk Assessment Guide (October 2003). It can be found at www.faa.gov.

Next, think through your recent flying experiences and make a note of the lowest weather conditions that you have comfortably experienced as a pilot in your VFR and, if applicable, IFR flying in the last 6–12 months. You might want to use the charts in *Figures 8-3* through *8-5* as guides for this assessment, but do not think that you need to fill in every square. In fact, you may not have, or even need, an entry for every category. Suppose that most of your flying takes place in a part of the country where clear skies and visibilities of 30 plus miles are normal. Your entry might specify the lowest VFR ceiling as 7,000, and the lowest visibility as 15 miles. You may have never experienced MVFR conditions at all, so you would leave those boxes blank.

For example, in a part of the country where normal summer flying often involves hazy conditions over relatively flat terrain, pilots who know the local terrain could regularly operate in hazy daytime MVFR conditions (e.g., 2,500 and four miles), and would use the MVFR column to record these values.

Category	Ceiling		Visibility
Visual Flight Rules VFR (green sky symbol)	Greater than 3,000 feet AGL	and	Greater than 5 miles
Marginal Visual Flight Rules MVFR (blue sky symbol)	1,000 to 3,000 feet AGL	and/or	3 to 5 miles
Instrument Flight Rules IFR (red sky symbol)	500 to below 1,000 feet AGL	and/or	1 mile to less than 3 miles
Low Instrument Flight Rules LIFR (magenta sky symbol)	below 500 feet AGL	and/or	less than 1 mile

Figure 8-1. *The regulations define weather flight conditions for visual flight rules (VFR) and instrument flight rules (IFR) in terms of specific values for ceiling and visibility.*

Certification, Training, and Experience Summary	
Certification Level	
Certificate level (e.g., private, commercial, ATP)	
Ratings (e.g., instrument, multiengine)	
Endorsements (e.g., complex, high performance, high altitude)	
Training Summary	
Flight review (e.g., certificate, rating, wings)	
Instrument Proficiency Check	
Time since checkout in airplane 1	
Time since checkout in airplane 2	
Time since checkout in airplane 3	
Variation in equipment (e.g., GPS navigators, autopilot)	
Experience	
Total flying time	
Years of flying experience	
Recent Experience (last 12 months)	
Hours	
Hours in this airplane (or identical model)	
Landings	
Night hours	
Night landings	
Hours flown in high density altitude	
Hours flown in mountainous terrain	
Crosswind landings	
IFR hours	
IMC hours (actual conditions)	
Approaches (actual or simulated)	

Figure 8-2. *Certification, training, and experience summary.*

Even in your home airspace, you should not consider flying down to VFR minimums at night—much less in the range of conditions defined as MVFR. For night VFR, anything less than a ceiling of at least 5,000, and visibility of at least seven to eight miles should raise a red flag.

Figure 8-3 shows how your entries would look in the Experience & Comfort Level Assessment VFR & MFR chart.

Experience and "Comfort Level" Assessment VFR & MVFR		
Weather Condition	**VFR**	**MVFR**
Ceiling	> 3,000	1,000–3,000
Day	—	2,500
Night	5,000	—
Visibility	> 5 miles	3–5 miles
Day	—	4 miles
Night	8 miles	—

Figure 8-3. *Experience and comfort level assessment for VFR and MVFR.*

If you fly IFR, the next part of the exercise shown in *Figure 8-4* is to record the lowest IFR conditions that you have comfortably, recently, and regularly experienced in your flying career. Again, be honest in your assessment. Although you may have successfully flown in low IFR (LIFR) conditions--down to a 300 foot ceiling and ¾ mile visibility—it does not mean you were "comfortable" in these conditions. Therefore, leave the LIFR boxes blank with entries for known "comfort level" in instrument meteorological conditions (IMC), as shown in *Figure 8-4*.

Experience and "Comfort Level" Assessment IFR & LIFR		
Weather Condition	**IFR**	**LIFR**
Ceiling	500–999	< 500
Day	800	—
Night	999	—
Visibility	1–3 miles	< 1 mile
Day	1 mile	—
Night	3 miles	—

Figure 8-4. *Experience and comfort level assessment for IFR and LIFR.*

If entries are combined into a single chart, the summary of your personal known "comfort level" for VFR, MVFR, IFR, and LIFR weather conditions would appear as shown in *Figure 8-5*.

Experience and "Comfort Level" Assessment Combined VFR & IFR				
Weather Condition	VFR	MVFR	IFR	LIFR
Ceiling				
Day		2,500		800
Night		5,000		999
Visibility				
Day		4 miles		1 mile
Night		8 miles		3 miles

Figure 8-5. *Experience and comfort level assessment for combined VFR and IFR.*

Step 3—Consider Other Conditions

Ceiling and visibility are the most obvious conditions to consider in setting personal minimums, but it is also a good idea to have personal minimums for wind and turbulence. As with ceiling and visibility, the goal in this step is to record the most challenging wind conditions you have comfortably experienced in the last 6–12 months—not necessarily the most challenging wind conditions you have managed to survive without bending an airplane. As shown in *Figure 8-6,* you can record these values for category and class, for specific make and model, or perhaps both.

Experience and "Comfort Level" Assessment Wind & Turbulence			
	SE	ME	Make/ Model
Turbulence			
Surface wind speed	10 knots	15 knots	
Surface wind gusts	5 knots	8 knots	
Crosswind component	7	7	

Figure 8-6. *Experience and comfort level assessment for wind and turbulence.*

In addition to winds, your "comfort level" inventory should also include factors related to aircraft performance. There are many variables, but start by completing the chart with reference to the aircraft and terrain most typical for the kind of flying you do most. *[Figure 8-7]* Remember that you want to establish a safety buffer, so be honest with yourself. If you have never operated to/from a runway shorter than 5,000 feet, the "shortest runway" box should say 5,000 feet. We will talk more about safe ways to extend personal minimums a bit later.

Experience and "Comfort Level" Assessment Performance Factors			
	SE	ME	Make/ Model
Performance			
Shortest runway	2,500	4,500	
Highest terrain	6,000	3,000	
Highest density altitude	3,000	3,000	

Figure 8-7. *Experience and comfort level assessment for performance factors.*

Step 4—Assemble and Evaluate

Now you have some useful numbers to use in establishing baseline personal minimums. Combining these numbers, the Baseline Personal Minimums chart in *Figure 8-8* shows how the whole picture might look.

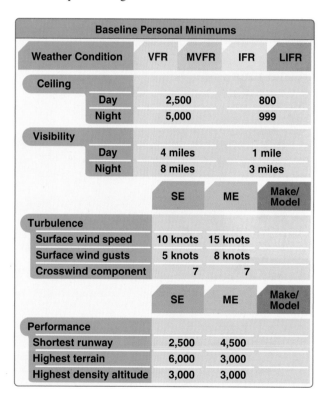

Baseline Personal Minimums				
Weather Condition	VFR	MVFR	IFR	LIFR
Ceiling				
Day		2,500		800
Night		5,000		999
Visibility				
Day		4 miles		1 mile
Night		8 miles		3 miles
		SE	ME	Make/ Model
Turbulence				
Surface wind speed		10 knots	15 knots	
Surface wind gusts		5 knots	8 knots	
Crosswind component		7	7	
		SE	ME	Make/ Model
Performance				
Shortest runway		2,500	4,500	
Highest terrain		6,000	3,000	
Highest density altitude		3,000	3,000	

Figure 8-8. *Baseline personal minimums.*

Step 5—Adjust for Specific Conditions

Any flight you make involves almost infinite combinations of pilot skill, experience, condition, and proficiency; aircraft equipment and performance; environmental conditions; and external influences. Both individually and in combination, these factors can compress the safety buffer provided by your baseline personal minimums. Consequently, you need a practical way to adjust your baseline personal minimums to accommodate specific conditions.

Note that the suggested adjustment factors are just that—a suggestion. If your flying experience is limited or if you do not fly very often, you might want to double these values. In addition, if your situation involves more than one special condition from the chart above, you will probably want to add the adjustment factor for each one. For example, suppose you are planning a night cross-country to an unfamiliar airport, departing after a full workday. If you decide to make this trip—or you might decide that it is safest to wait until the next day—the chart in *Figure 8-9* suggests that you should at least raise your baseline personal minimums by adding 1,000 feet to your ceiling value; one mile to visibility, and 1,000 feet to required runway length.

How about adjustments in the other direction? Some pilots fear that establishing personal minimums is a once and-for-all exercise. With time and experience, though, you can modify personal minimums to match growing skill and judgment. When you have comfortably flown to your baseline personal minimums for several months, you might want to sit down and assess whether and how to push the envelope safely. If, for instance, your personal minimums call for daytime visibility of at least five miles, and you have developed some solid experience flying in those conditions, you might consider lowering the visibility value to four miles for your next flight.

There are two important cautions:

1. Never adjust personal minimums to a lower value for a specific flight. The time to consider adjustments is when you are not under any pressure to fly, and when you have the time and objectivity to think honestly about your skill, performance, and comfort level during last the few flights. Changing personal minimums "on the fly" defeats the purpose of having them in the first place.

2. Keep all other variables constant. For example, if your goal is to lower your baseline personal minimums for visibility, don't try to lower the ceiling, wind, or other values at the same time. In addition, you never want to push the baseline if there are special conditions (e.g., unfamiliar aircraft, pilot fatigue) present for this flight. You might find it helpful to talk through both your newly established personal minimums and any "push-the-envelope" plans with a well-qualified flight instructor.

Step 6—Stick to the Plan!

Once you have done all the thinking required to establish baseline personal minimums, all you need to do next is stick to the plan. As most pilots know, that task is a lot harder than it sounds, especially when the flight is for a trip that you really want to make, or when you are staring into the faces of your disappointed passengers. Here's where personal minimums can be an especially valuable tool. Professional pilots live by the numbers, and so should you. Pre-established hard numbers can make it a lot easier to make a smart no go or divert decision than a vague sense that you can "probably" deal with the conditions that you are facing at any given time. In addition, a written set of personal minimums can also make it easier to explain tough decisions to passengers who are, after all, trusting their lives to your aeronautical skill and judgment.

	If you are facing		Adjust baseline personal minimums by
Pilot	Illness, use of medication, stress, or fatigue; lack of currency (e.g., have not flown for several weeks)	**Add**	At least 500 feet to ceiling
Aircraft	An unfamiliar airplane or an aircraft with unfamiliar avionics or other equipment		At least 1/2 mile to visibility
enVironment	Unfamiliar airports and airspace; different terrain or other unfamiliar characteristics	**Subtract**	At least 500 feet to runway length
External Pressures	"Must meet" deadlines, pressures from passengers, etc.		At least 5 knots from winds

Figure 8-9. *Examples of baseline personal mimimums.*

Chapter Summary

General aviation pilots enjoy a level of responsibility and freedom unique in aviation. Unlike the air carrier, corporate, and military communities, most GA pilots are free to fly when and where they choose. They are unencumbered by the strict regulatory structure that governs many other flight operations. However, the GA pilot is not supported by a staff of dispatchers and meteorologists, or governed by rigid operational guidelines designed to reduce risk. Pilots should not be lulled into a false sense of security simply because they are in compliance with the regulations. Judgment and aeronautical decision-making serve as the bridge between regulatory compliance and safety. Deciding if or when to undertake any flight lies solely with the pilot in command (PIC). GA pilots should remember that FAA regulations designed to prevent accidents and incidents come out AFTER the accident or incident.

A copy of the charts used in this chapter can be found in Appendix B. Pilots are encouraged to make a copy of this appendix, complete applicable charts, and use them prior to each flight.

Appendix A
Personal Assessment and Minimums

Each pilot should establish personal weather minimums, which may be (and often are) above FAA legal minimums for a VFR or IFR flight. Airlines and corporate flight departments set individual minimums above FAA weather minimums where pilot experience is limited. They also have operational minimums that apply when the aircraft has less than full operation of all systems for dealing with weather. Similarly, pilot personal minimums should be based on a clear assessment of pilot certification, experience, proficiency, and currency.

The assessment form below should be used to evaluate personal experience and qualifications. When the pilot obtains a new rating or upgrades a certificate, or when current experience level changes, the self-assessment factors should be reviewed (at least annually). If appropriate, revisions would then be made to the personal minimums. A copy of the appropriate personal minimums checklist should be carried with the pilot when making important risk management decisions.

Certification, Training, and Experience Summary	Self-Assessment Factors	Revised Self-Assessment
Certification/ratings (e.g., private, multi-engine; instrument)		
Highest certificate level 7 ratings (including complex aircraft)		
Training		
Flight review (e.g., certificate, rating, Wings Program completion)		
Instrument Proficiency Check		
Time since checkout in aircraft #1:		
Time since checkout in aircraft #2:		
Time since checkout in aircraft #3:		
Variation in equipment (GPS navigator), number of different panels		
Experience		
Total flying time in hours		
Number of years flying		

Certification, Training, and Experience Summary	Self-Assessment Factors	Revised Self-Assessment
Hours in the last year		
Hours in this or identical airplane in last year		
Landings in last year		
Night hours in last year		
Night landings in last year		
High density altitude hours in last year		
Mountainous terrain hours in last year		
Strong crosswind or gusty landings in last year		
IFR hours in last year		
Actual hours in IMC in the last year		
Approaches (actual or simulated) in last year		

Personal Minimums
VER Pilot

Cut and Fold	Cut and Fold	Cut and Fold
Condition	Example: 100 Hour VFR Pilot	Your Personal Minimums
Minimum visibility – day VFR	5 miles	
Minimum visibility – night VFR	7 miles	
Minimum ceiling – day VFR	3,000 feet	
Minimum ceiling – night VFR	5,000 feet	
Surface wind speed & gusts	15 knots 5 knot gust	
Maximum cross wind	7 knots	
Other VFR (e.g., mountain flying, over water beyond gliding distance)	Consult instructor/mentor	
Fuel reserves (day VFR)	1 hour	
Fuel reserves (night VFR)	1½ hour	

Personal Minimums
IFR Pilot

Cut and Fold	Cut and Fold	Cut and Fold
Condition	Example: 300 Hour IFR Pilot	Your Personal Minimums
Minimum visibility – day VFR	3 miles	
Minimum visibility – night VFR	5 miles	
Minimum ceiling – day VFR	2,000 feet	
Minimum ceiling – night VFR	3,000 feet	
Surface wind speed & gusts	15 knots 5 knot gust	
Maximum cross wind	7 knots	
IFR approach ceiling	Minimums + 500 feet	
IFR approach visibility (precision approaches)	Minimums + ½ mile	
IFR approach visibility (nonprecision approaches)	Minimums + 1 mile	
Other IFR (e.g., icing)	Consult instructor/mentor	
Fuel reserves (day VFR)	1 hour	
Fuel reserves (night or IFR) 1½ hour	1½ hour	

Appendix B

Sample Risk Management Scenarios

As a part of their responsibilities, flight instructors should include training on evaluating risk and using tools to manage risks including making go/no-go flight decisions. Once a pilot leaves the instructor's supervision, that pilot makes those decisions for him or herself.

The following scenarios are designed to provide a discussion platform that can be used to apply risk management procedures and checklists such as the PAVE checklists. Instructors are encouraged to add scenarios from personal experience or adapt other familiar scenarios. The answers and explanations to the questions for each of the following scenarios can be found at the end of this appendix.

VFR Scenarios

Scenario 1

You are a 32 year old, 325-hour, non-instrument-rated private pilot. You have about 75 hours on long cross-country flights including one less than 3 weeks ago. You and your wife are planning to leave after work for a 400 NM flight to attend your wife's best friend's 11:00 AM wedding the next day. You will take off about 30 minutes before sunset.

The current weather is good and the forecast calls for clear skies, good visibility, and negligible wind en route and at your destination. You have planned to cruise at 9,500 because about 250 miles of the route is over rugged terrain (few airports) that goes up to 7,500 feet. Your aircraft is a typical general aviation single, 160 horsepower (HP), 600-mile range, and no oxygen is available.

You have decided that the combination of fatigue, rough terrain, and night make this flight, as originally planned, undesirable.

Which of the potential solution(s) below would best manage the risk of this trip?

    a.   Rearrange your work schedule to leave early and land in daylight.

    b.   Get an airline reservation for your wife and delay your departure until tomorrow morning.

    c.   Postpone the trip until tomorrow morning.

Scenario 2

It is early October, you have 135 hours in your logbook, and this is your first trip to see your mom since getting your license last summer. It will also be your longest trip—a little over 700 miles. You just took a 150-mile cross-country two weeks ago.

You recently asked your instructor to help you make up a personal SOP. You both agreed on the following rules: fuel reserves—1 hour day, 1½ hours night; ceiling—2,000 feet above your planned en route altitude; visibility—5 miles; a maximum work/duty day of 12 hours, with no more than 8 hours flying time; a minimum of 6 hours sleep.

You got an outlook weather briefing the night before, and the weather should be pretty good. You are making this trip to be at your Mom's 60th birthday party, but you told her that you would not come unless it's completely safe, and she understands. Your sister, Joan, will be there to help her celebrate, and you will see her when you are at Joan's on Halloween—that's less than a month away.

You told your Mom that you would give her a call after you have landed at your hometown. By the time Mom and Joan get to the airport, you will be done post-flighting, fueling, and tying down the plane. You will be ready to go. The forecast at hometown calls for 2,500 scattered and visibility greater than 6 miles. The weather briefer told you that the minimum conditions en route would be 6,500 broken and 5 miles visibility. You are planning to cruise at 4,500, but you could go lower, except for that 100-mile section over the hills.

Allowing for the fuel and sandwich stop, you should get there about 4:00 PM. You expect to land with 1 hour and 20 minutes of fuel left in the tanks at the fuel stop, and 1 hour and 10 minutes left at your destination. Unfortunately, the airplane you are most comfortable with, 87EV, had to go into the shop. You have reserved the FBO's newest plane, 4892M, and your instructor told you that the only difference between it and 87EV is the avionics manufacturer. He also told you that he would make sure that you understood the avionics before you left.

You had a good night's sleep, and you are feeling great. You have reviewed the **IMSAFE** checklist, and you are ready to go.

Considering the PAVE checklist, which, if any, risk category factors are marginal for this flight?

    a. Pilot;

    b. Aircraft;

    c. enVironment;

    d. External Pressures

Based on these conditions, would you go on his trip, or not?

Scenario 3

You are a 350-hour non-instrument private pilot, and you are 3 hours into a 4-hour VFR cross-country. There is a stronger headwind than forecast, and while you left with 5 hours fuel, you think you are seeing a higher than normal fuel burn.

You are tired and hungry, and your wife has mentioned a need for a bathroom, and she is not about to use one of those plastic things you carry in your flight bag.

The weather at your destination is still VFR, but the temperature and dew point are closing as darkness falls. Your destination airport lies on the eastern edge of a very large lake and the winds are from the west.

You are thinking through the consequences of these issues and go back to your **PAVE** checklist.

    Pilot

    Aircraft

    en**V**ironment

    External pressures

Based on this information, and considering the PAVE risk factor category(ies) that are marginal, should you continue on to your destination, or should you land at a suitable airport?

IFR Scenarios

Scenario 1

You are an IFR-rated private pilot with 850 hours, 50 of which are in actual IMC. You have planned a 325-mile trip to meet with an important client. The destination forecast is for IFR ceilings and visibilities with conditions tending to worsen at your ETA. Because of that, you have chosen a suitable alternate 100 miles distant, where the weather is forecast to be better.

You have performed the IMSAFE checklist and, while there is some stress associated with the business meeting, you are in fine shape, mentally and physically. While you are technically IFR current, the last approaches you flew were 4½ months ago.

This is your first trip to today's destination. You are looking forward to the flight so that you can check out your new IFR GPS. The only approach to your destination is a GPS stand-alone, and although you've never flown one, you have spent about an hour of GPS practice on a PCATD. Your client just called to tell you that it is critical that you make the meeting—he is leaving town that night.

Which choice from the list below reflects the best preflight risk management for this situation?

a. You take off as scheduled, having made arrangements at the alternate to have a rental car ready should you need to land there and hope the weather holds.

b. You rush your plans, leave early, and try to beat the worsening weather by arriving earlier.

c. You decide that today is not the day to fly yourself to this destination. You move the meeting to later in the afternoon and catch an airline flight.

Scenario 2

You are an 845-hour private pilot with an instrument rating. You are planning a 475-mile winter trip with your wife and teenage son to visit relatives for the 3-day holiday weekend. You have recently purchased a 180 HP Cessna 172 that sports a service ceiling of 14,000 feet.

You have flown over the mountains once several years ago. The most direct route to your destination requires 13,500 feet to clear the mountain range by at least 1,000 ft. and the nearest airway Minimum Enroute Altitude is 14,000 feet. In spite of the fact that you really want to justify the utility of your airplane to your wife, you begin to doubt that your 172 is the right airplane for the trip.

Which of the options from the list below would make the aircraft risk factor acceptable for this trip?

a. C-172. Fly the 172, but revise your schedule to allow more travel time and change your route to one with a 10,000-foot maximum airway MEA.

b. PA-32. Schedule the FBO's Turbo Saratoga that you flew a couple of years ago, and get refresher checkout.

c. C-172. Fly the 172 on the original route, but plan a stop at a 7,000-foot airport on this side of the highest ridges to check out conditions and get local advice. Take portable oxygen.

Scenario 3

You are an instrument-rated private pilot and have logged 942 hours since you started flying 5 years ago. You regularly fly your own, well-equipped Bonanza to see your company's customers within a 700-mile radius.

You are planning a 365-mile trip with Jim, your chief engineer, to visit a long-time client. You need to figure out what is going wrong with the newly installed framis plate—your company's newly upgraded flagship product. The client has just landed a significant government contract, but their new framis plate is holding up production. The client is so upset, that he is threatening to go to your competitor. You and Jim were at the office until 2:30 AM going over the calculations, and think you have a solution. You will only need a few hours to install the changes.

You have logged over 6 approaches and 12 hours actual instrument time during the last 2 months including three approaches with weather right at minimums. You also completed a Bonanza type refresher course less than a year ago.

The forecast for the destination airport is ¾-mile visibility in rain and 300-foot overcast, with temporary conditions of ¼ mile in rain and fog and 100-foot overcast. The ILS approach decision height is 200 feet and the visibility minimum is ½ mile. There are no thunderstorms observed or forecast and the freezing level is 2,000 feet above your filed altitude. These conditions are widespread and the best alternate is another 100 miles beyond the destination. The weather at the alternate just makes the legal minimums of a 600-foot ceiling and 2 miles visibility to qualify it as an alternate.

You estimate that you will have 1 hour of fuel at the alternate. The avionics suite in your Bonanza includes dual Nav/Com and an IFR GPS. After reading the DUAT printout, you review your **PAVE** checklist.

Pilot

Aircraft

en**V**ironment

External pressures

Considering the following potential risk factors.

A. Pilot proficiency

B. Pressure to make the trip

C. Thunderstorms or icing

D. Aircraft performance

E. Ceiling and visibility

F. Avionics

G. Fatigue

H. Stress

Which, if any, of the potential risk factors would cause you to rate any of the PAVE risk categories as marginal?

Based on your PAVE checklist, should you go, or not go, on this trip?

Scenario 4

You are a 2,000-hour instrument rated pilot flying turbo-charged, complex single to a three-day seminar you're conducting. After departing a mid-point fuel stop for the final 2-½ hour leg, and climbing to VFR conditions on top of an overcast, the generator fails. The destination weather is forecast for overcast clouds at 1,000 feet and 3 miles visibility. You expect to arrive at your destination shortly before sunset.

Using your checklist, you accept the changed reality of this failure. Consider the following possible alternatives and choose the alternative(s) that would be acceptable ways to deal this change:

a. Shut down all the electrical equipment and dead reckon to the destination. Over the destination, turn the master on and one NAV/COM for the approach.

b. Declare an emergency with ATC, reduce electrical load, return and land at your fuel stop.

c. Advise ATC of the problem, shut down all electrical equipment, and dead reckon to an area of known clear weather.

Answers and Explanations to Scenario Questions
VFR Scenario 1

a. Rearrange your work schedule to leave early and land in daylight.

Leaving early to land in daylight reduces the performance level risk from fatigue and eliminates the night risk factor.

b. Get an airline reservation for your wife, and delay your departure until tomorrow morning.

If you put your wife on an airline flight so she can be sure to make the wedding, you will reduce the pressure to make the flight the next day if there are uncomfortable levels of other risk at that time.

c. Postpone the trip until tomorrow morning.

Postponing the trip until tomorrow morning reduces the performance level risk from fatigue and eliminates the night risk factors, but leaves no other option for making the wedding on time. You are vulnerable to the pressure to complete the flight even if other risk factors, such as weather, go above normal.

VFR Scenario 2

The only marginal **PAVE** checklist item is:

3. enVironment

At a cruising altitude of 4,500 feet (required over the hills), the forecast en route ceiling of 6,500 exactly equals the pilot's personal minimum of 2,000 feet above his cruising altitude (cloud heights in Area Forecasts are MSL unless denoted AGL or CIG). Likewise, the forecast en route visibility of 5 miles exactly equals the pilot's personal minimum. There is no room for the weather to deteriorate without going below the pilot's personal minimums, so the weather is a marginal item in the enVironment category.

PAVE checklist items that are not marginal:

a. **P**ilot

You feel well, and you have satisfactorily completed the I'M SAFE checklist. You have recently completed your checkride and have 135 hours. You have also flown a cross-country flight within the last 2 weeks. You are qualified for this flight.

b. **A**ircraft

This airplane is the same model airplane with which you are familiar, and you will get instruction on the avionics differences.

c. External Pressures

You have worked out alternatives regarding seeing your mother, and no one will be meeting you until you call them on arrival.

Go/No-go decision following review of **PAVE** checklist:

Go

With only one risk factor marginal on your PAVE checklist, under most circumstances, you could comfortably decide to make this flight.

VFR Scenario 3

The marginal **PAVE** checklist items are:

a. **P**ilot

You are hungry, fatigued, and feeling the stress from your wife's discomfort. These make the Pilot risk category marginal.

b. **A**ircraft

You originally planned a 1-hour fuel reserve (5 hours of fuel for a planned 4-hour trip), but the headwind is stronger than forecast and your fuel burn appears to be higher than normal. Since you will no longer have the 1-hour fuel reserve you planned, the Aircraft risk category is marginal.

c. enVironment

The airport location on the downwind side of a large lake and the closing temperature-dew point spread makes it likely that fog will form. This makes the enVironment category marginal.

The **PAVE** checklist item that is not marginal:

4. External Pressures

    There is no indication of any additional external pressure to reach the destination other than your natural inclination to complete your planned trip.

Continue/Land decision following review of **PAVE** checklist:

Land

You should take action any time you have two or more **PAVE** risk categories that are marginal. If you are airborne, make the decision to land. In this case, the **P**ilot (hunger, fatigue, and stress from spouse), the **A**ircraft (reduced fuel reserve), and the en**V**ironment (closing temperature-dew point spread) risk factors are all marginal. There is no indication that the External Pressure risk factor is marginal.

IFR Scenario 1

a. You take off as scheduled, having made arrangements at the alternate airport to have a rental car ready should you need to land there, and hope the weather holds.

   All the risk factors are still present, making the **P**ilot, en**V**ironment, and **E**xternal Pressures all marginal. Once airborne, your goal-orientated nature will pressure you to attempt to land at your destination (such as pressing minimums on the approach).

b. You rush your plans, leave early, and try to beat the worsening weather by arriving earlier.

   By rushing, you may take shortcuts and miss something in your planning, ground or inflight procedures, and you have now added the additional stress of trying to beat the weather.

c. You decide that today is not the day to fly yourself to this destination. You move the meeting to later in the afternoon and catch an airline flight.

With marginal instrument proficiency and no in-cockpit experience flying GPS approaches, you have substantially increased risk in the Pilot category for this flight in additional to marginal risk factors in both the en**V**ironment and External Pressures categories.

IFR Scenario 2

a. C-172. Fly the 172, but revise your schedule to allow more travel time and change your route to one with a 10,000-foot maximum airway MEA.

   Flying a route well within the capabilities of your airplane reduces the Aircraft risk factor.

b. PA-32. Schedule the FBO's Turbo Saratoga that you flew a couple of years ago, and get refresher checkout.

   Flying an airplane, such as a Turbo Saratoga that is capable of comfortably cruising at the airway Minimum Enroute Altitude or higher reduces the Aircraft risk factor.

c. C-172. Fly the 172 on the original route, but plan a stop at a 7,000-foot airport on this side of the highest ridges to check out conditions and get local advice. Take portable oxygen.

   The stop, getting local advice, and carrying oxygen do nothing about the risk of flying the 172 at its performance limit.

IFR Scenario 3

The marginal **PAVE** checklist items are:

a.   **P**ilot

You are a current and proficient instrument pilot, but you have had less than 4 hours sleep, and you are stressed about your client's malfunctioning product. The Pilot risk category is marginal.

c.   en**V**ironment

There are no forecast thunderstorms and no mention of icing at you planned flight altitude, and the freezing level is 2,000 feet above it (freezing level forecasts are pretty dependable). However, the ceiling and visibility at the destination are generally expected to be slightly above the ILS approach minimums, there will be periods when the weather is expected to be lower than the minimums. The enVironment risk category is marginal.

d.   **E**xternal Pressures

You have a very strong motivation to make this trip, and that makes the External Pressures risk category marginal.

The **PAVE** checklist item that is not marginal:

b.   **A**ircraft

It appears that this trip is comfortably within the capabilities of your airplane including the avionics suite.

Go/No-go decision following review of **PAVE** checklist:

No-Go

The **PAVE** checklist shows more than two risk categories as marginal leading to an insidious cumulative effect. If you have marginal items in two or more categories, do not go.

IFR Scenario 4

a.   Shut down all the electrical equipment and dead reckon to the destination. Over the destination, turn the master on and one NAV/COM for the approach.

Reality is that the airplane has changed, and you need to come to terms with these changes. You must change your plans. There are many possible consequences of continuing to the destination including worsening weather and a significant possibility that you might have no battery power by the time you get there. Since your planned arrival is just before sunset, any delay will mean that you will arrive in the dark.

b.   Declare an emergency with ATC, reduce electrical load, return and land at your fuel stop.

You would get ATC's full attention by declaring an emergency and have a better chance of landing with some electrical power.

c.   Advise ATC of the problem, shut down all electrical equipment, and dead reckon to an area of known clear weather.

If you have good information on clear weather areas and plenty of fuel to get there, dead reckoning to such an area is a good alternative.

Appendix C

Flight Safety Foundation
CFIT Checklist
Evaluate the Risk and Take Action

Flight Safety Foundation (FSF) designed this controlled-flight-into-terrain (CFIT) risk-assessment safety tool as part of its international program to reduce CFIT accidents, which present the greatest risks to aircraft, crews and passengers. The FSF CFIT Checklist is likely to undergo further developments, but the Foundation believes that the checklist is sufficiently developed to warrant distribution to the worldwide aviation community.

Use the checklist to evaluate specific flight operations and to enhance pilot awareness of the CFIT risk. The checklist is divided into three parts. In each part, numerical values are assigned to a variety of factors that the pilot/operator will use to score his/her own situation and to calculate a numerical total.

In Part I: CFIT Risk Assessment, the level of CFIT risk is calculated for each flight, sector or leg. In Part II: CFIT Risk-reduction Factors, Company Culture, Flight Standards, Hazard Awareness and Training, and Aircraft Equipment are factors, which are calculated in separate sections. In Part III: Your CFIT Risk, the totals of the four sections in Part II are combined into a single value (a positive number) and compared with the total (a negative number) in Part I: CFIT Risk Assessment to determine your CFIT Risk Score. To score the checklist, use a nonpermanent marker (do not use a ball point pen or pencil) and erase with a soft cloth.

Part I: CFIT Risk Assessment

Section 1—Destination CFIT Risk Factors	Value	Score
Airport and Approach Control Capabilities:		
ATC approach radar with MSAWS	0	_____
ATC minimum radar vectoring charts	0	_____
ATC radar only	−10	_____
ATC radar coverage limited by terrain masking	−15	_____
No radar coverage available (out of service/not installed)	−30	_____
No ATC service	−30	_____
Expected Approach:		
Airport located in or near mountainous terrain	−20	_____
ILS	0	_____
VOR/DME	−15	_____
Nonprecision approach with the approach slope from the FAF to the airport TD shallower than 2 3/4 degrees	−20	_____
NDB	−30	_____
Visual night "black-hole" approach	−30	_____
Runway Lighting:		
Complete approach lighting system	0	_____
Limited lighting system	−30	_____

Controller / Pilot Language Skills:

Controllers and pilots speak different primary languages.. –20   _____

Controllers' spoken English or ICAO phraseology poor .... –20   _____

Pilots' spoken English poor.. –20   _____

Departure:

No published departure procedure .... –10   _____

Destination CFIT Risk Factors Total   (–) _____

Section 2—Risk Multiplier   Value   Score

Your company's Type of Operation (select single highest applicable value):

Scheduled .... 1.0   _____

Nonscheduled .... 1.2   _____

Corporate.. 1.3   _____

Charter.. 1.5   _____

Business owner/pilot .... 2.0   _____

Regional .... 2.0   _____

Freight .... 2.5   _____

Domestic... 1.0   _____

International... 3.0   _____

Departure/Arrival Airport (select single highest applicable value):

Australia/New Zealand.. 1.0   _____

United Stated/Canada.. 1.0   _____

Western Europe .... 1.3   _____

Middle East .... 1.1   _____

Southeast Asia .... 3.0   _____

Euro-Asia (Eastern Europe and Commonwealth of Independent States) .... 3.0   _____

South America/Caribbean .... 5.0   _____

Africa.. 8.0   _____

Weather/Night Conditions (select single highest applicable value):

Night—no moon.. 2.0   _____

IMC .... 3.0   _____

Night and IMC .... 5.0   _____

Crew (select only one value):

Single-pilot flight crew... 1.5   _____

Flight crew duty day at maximum and ending with a night nonprecision approach.................... 1.2   _____

Flight crew crosses five or more time zones .... 1.2   _____

Third day of multiple time-zone crossings... 1.2   _____

Add Multiplier Values to Calculate Risk Multiplier Total   _____

Destination CFIT Risk Factors Total × Risk Multiplier Total = CFIT Risk Factors Total   (–) _____

Part II: CFIT Risk-Reduction Factors

Section 1—Company Culture	Value	Score

Corporate/company management:

	Value	Score
Places safety before schedule	20	_____
CEO signs off on flight operations manual	20	_____
Maintains a centralized safety function	20	_____
Fosters reporting of all CFIT incidents without threat of discipline	20	_____
Fosters communication of hazards to others	15	_____
Requires standards for IFR currency and CRM training	15	_____
Places no negative connotation on a diversion or missed approach	20	_____

115–130 points	Tops in company culture	
105–115 points	Good, but not the best	**Company Culture Total** (+)_____
80–105 points	Improvement needed	
Less than 80 points	High CFIT risk	

Section 2—Flight Standards	Value	Score

Specific procedures are written for:

	Value	Score
Reviewing approach or departure procedures charts	10	_____
Reviewing significant terrain along intended approach or departure course	20	_____
Maximizing the use of ATC radar monitoring	10	_____
Ensuring pilot(s) understand that ATC is using radar or radar coverage exists	10	_____
Altitude changes	10	_____
Ensuring checklist is complete before initiation of approach	10	_____
Abbreviated checklist for missed approach	10	_____
Briefing and observing MSA circles on approach charts as part of plate review	10	_____
Checking crossing altitudes at IAF positions	10	_____
Checking crossing altitudes at FAF and glideslope centering	10	_____
Independent verification by PNF of minimum altitude during stepdown DME (VOR/DME or LOC/DME) approach	20	_____
Requiring approach/departure procedure charts with terrain in color, shaded contour formats	20	_____
Radio-altitude setting and light-aural (below MDA) for backup on approach	10	_____
Independent charts for both pilots, with adequate lighting and holders	10	_____
Use of 500-foot altitude call and other enhanced procedures for NPA	10	_____
Ensuring a sterile (free from distraction) cockpit, especially during IMC/night approach or departure	10	_____
Crew rest, duty times and other considerations especially for multiple-time-zone operations	20	_____
Periodic third-party or independent audit of procedures	10	_____
Route and familiarization checks for new pilots		
Domestic	10	_____
International	20	_____
Airport familiarization aids, such as audiovisual aids	10	_____

Section 2—Flight Standards (continued)	Value	Score
First officer to fly night or IMC approaches and the captain to monitor the approach...	20	_____
Jump-seat pilot (or engineer or mechanic) to help monitor terrain clearance and the approach in IMC or night conditions.......................................	20	_____
Insisting that you fly the way that you train	25	_____

		Flight Standards Total	(+)_____
300–335 points	Tops in CFIT flight standards		
270–300 points	Good, but not the best		
200–270 points	Improvement needed		
Less than 200 points	High CFIT risk		

Section 3—Hazard Awareness and Training	Value	Score
Your company reviews training with the training department or training contractor.................	10	_____
Your company's pilots are reviewed annually about the following:		
Flight standards operating procedures..	20	_____
Reasons for and examples of how the procedures can detect a CFIT "trap"	30	_____
Recent and past CFIT incidents/accidents...	50	_____
Audiovisual aids to illustrate CFIT traps	50	_____
Minimum altitude definitions for MORA, MOCA, MSA, MEA, etc.	15	_____
You have a trained flight safety officer who rides the jump seat occasionally	25	_____
You have flight safety periodicals that describe and analyze CFIT incidents...........	10	_____
You have an incident/exceedance review and reporting program	20	_____
Your organization investigates every instance in which minimum terrain clearance has been compromised..	20	_____
You annually practice recoveries with GPWS in the simulator...........................	40	_____
You train the way that you fly	25	_____

		Hazard Awareness and Training Total	(+)_____
285–315 points	Tops in CFIT training		
250–285 points	Good, but not the best		
190–250 points	Improvement needed		
Less than 190 points	High CFIT risk		

Section 4—Aircraft Equipment	Value	Score
Aircraft includes:		
Radio Altimeter with cockpit display of full 2,500-foot range—captain only	20	_____
Radio Altimeter with cockpit display of full 2,500-foot range—copilot...................	10	_____
First-generation GPWS	20	_____
Second-generation GPWS or better	30	_____
GPWS with all approved modifications, data tables and service bulletins to reduce false warnings	10	_____
Navigation display and FMS...	10	_____
Limited number of automated altitude callouts...	10	_____
Radio-altitude automated callouts for Nonprecision approach (not heard on ILS approach) and procedure	10	_____
Preselected radio altitudes to provide automated callouts that would not be heard during normal nonprecision approach	10	_____
Barometric altitudes and radio altitudes and radio altitudes to give automated "decision" or "minimums" callout	10	_____
An automated excessive "bank angle" callout	10	_____

Section 4—Aircraft Equipment (continued)	Value	Score
Auto flight/vertical speed model	10	_____
Auto flight/vertical speed mode with no GPWS	20	_____
GPS or other long-range navigation equipment to supplement		
NDB-only approach	15	_____
Terrain-navigation display	20	_____
Ground-mapping radar	10	_____

175–195 points	Excellent equipment to minimize CFIT risk		
155–175 points	Good, but not the best	**Aircraft Equipment Total**	(+)_____*
115–155 points	Improvement needed		
Less than 115 points	High CFIT risk		

Company Culture _____ **+ Flight Standards** _____ **+ Hazard Awareness and Training** _____
+ Aircraft Equipment _____ **= CFIT Risk-reduction Factors Total (+)** _____

* If any section in Part II scores less than "Good," thorough review is warranted of that aspect of the company's operation.

Part III: Your CFIT Risk

Part I CFIT Risk Factors Total (–) _____ + Part II CFIT Risk-Reduction Factors Total (+) _____
= CFIT Risk Score (+) _____

A negative CFIT Risk Score indicates a significant threat; review the sections in Part II and
determine what changes and improvements can be made to reduce CFIT risk

Glossary

14 CFR. See Title 14 of the Code of Federal Regulations.

Acceptable risk. That part of identified risk that is allowed to persist without further engineering or management action. Making this decision is a difficult yet necessary responsibility of the managing activity. This decision is made with full knowledge that it is the user who is exposed to this risk.

ADM. See aeronautical decision-making.

Aeronautical decision-making. A systematic approach to the mental process used consistently by pilots to determine the best course of action in response to a given set of circumstances. It is what a pilot intends to do based on the latest information he or she has.

Aerodynamics. The science of the action of air on an object, and with the motion of air on other gases. Aerodynamics deals with the production of lift by the aircraft, the relative wind, and the atmosphere.

Aircraft. A device that is used, or intended to be used, for flight.

A/FD. See Airport/Facility Directory.

Airplane Flight Manual (AFM). A document developed by the airplane manufacturer and approved by the Federal Aviation Administration (FAA). It is specific to a particular make and model airplane by serial number, and it contains operating procedures and limitations.

Airport/Facility Directory (A/FD). An FAA publication containing information on all airports, communications, and NAVAIDs.

ATC. Air Traffic Control.

Attitude management. The ability to recognize hazardous attitudes in oneself and the willingness to modify them as necessary through the application of an appropriate antidote thought.

Automated Surface Observing System (ASOS). Weather reporting system which provides surface observations every minute via digitized voice broadcasts and printed reports.

Automated Weather Observing System (AWOS). Automated weather reporting system consisting of various sensors, a processor, a computer-generated voice subsystem, and a transmitter to broadcast weather data.

Automatic terminal information service (ATIS). The continuous broadcast of recorded non-control information in selected terminal areas. Its purpose is to improve controller effectiveness and relieve frequency congestion by automating repetitive transmission of essential but routine information.

Autopilot. An automatic flight control system that keeps an aircraft in level flight or on a set course. Automatic pilots can be directed by the pilot, or they may be coupled to a radio navigation signal.

Aviation medical examiner (AME). A physician with training in aviation medicine designated by the Civil Aerospace Medical Institute (CAMI).

Aviation Routine Weather Report (METAR). Observation of current surface weather reported in a standard international format.

AWOS. See Automated Weather Observing System.

Checklist. A tool that is used as a human factors aid in aviation safety. It is a systematic and sequential list of all operations that must be performed to accomplish a task properly.

Controlled flight into terrain (CFIT). An accident whereby an airworthy aircraft, under pilot control, inadvertently flies into terrain, an obstacle, or water.

Course. The intended direction of flight in the horizontal plane measured in degrees from north.

Crew resource management (CRM). The application of team management concepts in the flight deck environment. It was initially known as cockpit resource management, but as CRM programs evolved to include cabin crews, maintenance personnel, and others, the phrase "crew resource management" was adopted. This includes single pilots, as in most general aviation aircraft. Pilots of small aircraft, as well as crews of larger aircraft, must make effective use of all available resources: human, hardware, and information. A current definition includes all groups routinely working with the flight crew who are involved in decisions required to operate a flight safely. These groups include, but are not limited to pilots, dispatchers, cabin crewmembers, maintenance personnel, and air traffic controllers. CRM is one way of addressing the challenge of optimizing the human/machine interface and accompanying interpersonal activities.

CRM. See crew resource management.

DA. See decision altitude.

Dead reckoning. Navigation of an airplane solely by means of computations based on airspeed, course, heading, wind direction and speed, groundspeed, and elapsed time.

Decision altitude (DA). A specified altitude in the precision approach, charted in feet MSL, at which a missed approach must be initiated if the required visual reference to continue the approach has not been established.

Decision height (DH). A specified altitude in the precision approach, charted in height above threshold elevation, at which a decision must be made either to continue the approach or to execute a missed approach.

DH. See decision height.

Direct User Access Terminal System (DUATS). A system that provides current FAA weather and flight plan filing services to certified civil pilots via personal computer, modem, or telephone access to the system. Pilots can request specific types of weather briefings and other pertinent data for planned flights.

DUATS. See direct user access terminal system.

EFAS. See En Route Flight Advisory Service.

EFD. See electronic flight display.

Electronic flight display (EFD). For the purpose of standardization, any flight instrument display that uses LCD or other image-producing system (cathode ray tube (CRT), etc.)

Emergency. A distress or urgent condition.

En Route Flight Advisory Service (EFAS). An en route weather-only AFSS service.

External pressures. Influences external to the flight that create a sense of pressure to complete a flight—often at the expense of safety.

FAA. Federal Aviation Administration.

Federal Aviation Administration (FAA). An agency of the United States Department of Transportation with authority to regulate and oversee all aspects of civil aviation in the United States.

Flight director indicator (FDI). One of the major components of a flight director system, it provides steering commands that the pilot (or the autopilot, if coupled) follows.

Flight level (FL). A measure of altitude (in hundreds of feet) used by aircraft flying above 18,000 feet with the altimeter set at 29.92 "Hg.

Flight management system (FMS). Provides pilot and crew with highly accurate and automatic long-range navigation capability, blending available inputs from long- and short-range sensors.

Flightpath. The line, course, or track along which an aircraft is flying or is intended to be flown.

FMS. See flight management system.

General aviation. All flights other than military and scheduled airline flights, both private and commercial.

GPS Landing System (GLS). An instrument approach with lateral and vertical guidance with integrity limits (similar to barometric vertical navigation (Baro VNAV).

Global Navigation Satellite System (GNSS). Satellite navigation system that provides autonomous geospatial positioning with global coverage. It allows small electronic receivers to determine their location (longitude, latitude, and altitude) to within a few meters using time signals transmitted along a line of sight by radio from satellites.

Global positioning system (GPS). Navigation system that uses satellite rather than ground-based transmitters for location information.

GLS. See GPS Landing System.

GNSS. See Global Navigation Satellite System.

GPS. See Global Positioning System.

Hazard. A present condition, event, object, or circumstance that could lead to or contribute to an unplanned or undesired event, such as an accident. It is a source of danger. For example, a nick in the propeller represents a hazard.

Hazardous attitudes. Five aeronautical decision-making attitudes that may contribute to poor pilot judgment: anti-authority, impulsivity, invulnerability, macho, and resignation.

Hazardous Inflight Weather Advisory Service (HIWAS). Service providing recorded weather forecasts broadcast to airborne pilots over selected VORs.

Human behavior. The product of factors that cause people to act in predictable ways.

Human factors. A multidisciplinary field encompassing the behavioral and social sciences, engineering, and physiology, to consider the variables that influence individual and crew performance for the purpose of optimizing human performance and reducing errors.

Hypoxia. A state of oxygen deficiency in the body sufficient to impair functions of the brain and other organs.

Identified risk. Risk that has been determined through various analysis techniques. The first task of system safety is to identify, within practical limitations, all possible risks.

IFR. See instrument flight rules.

IMC. See instrument meteorological conditions.

Instrument flight rules (IFR). Rules and regulations established by the Federal Aviation Administration to govern flight under conditions in which flight by outside visual reference is not safe. IFR flight depends upon flying by reference to instruments in the flight deck, and navigation is accomplished by reference to electronic signals.

Instrument landing system (ILS). An electronic system that provides both horizontal and vertical guidance to a specific runway, used to execute a precision instrument approach procedure.

Instrument meteorological conditions (IMC). Meteorological conditions expressed in terms of visibility, distance from clouds, and ceiling less than the minimums specified for visual meteorological conditions, requiring operations to be conducted under IFR.

Judgment. The mental process of recognizing and analyzing all pertinent information in a particular situation, a rational evaluation of alternative actions in response to it, and a timely decision on which action to take.

Mean sea level. The average height of the surface of the sea at a particular location for all stages of the tide over a 19-year period.

MFD. See multifunction display.

MSL. See mean sea level.

Multifunction display (MFD). Small screen (CRT or LCD) in an aircraft that can be used to display information to the pilot in numerous configurable ways. Often an MFD will be used in concert with a primary flight display.

National Transportation Safety Board (NTSB). A United States Government independent organization responsible for investigations of accidents involving aviation, highways, waterways, pipelines, and railroads in the United States. NTSB is charged by congress to investigate every civil aviation accident in the United States.

NAVAID. Navigational aid.

NM. Nautical mile.

NOTAM. See Notice to Airmen.

Notice to Airmen (NOTAM). A notice filed with an aviation authority to alert aircraft pilots of any hazards en route or at a specific location. The authority in turn provides means of disseminating relevant NOTAMs to pilots.

NTSB. See National Transportation Safety Board.

Optical illusion. A misleading visual image. For the purpose of this handbook, the term refers to the brain's misinterpretation of features on the ground associated with landing, which causes a pilot to misread the spatial relationships between the aircraft and the runway.

Orientation. Awareness of the position of the aircraft and of oneself in relation to a specific reference point.

Personality. The embodiment of personal traits and characteristics of an individual that are set at a very early age and extremely resistant to change.

PFD. See primary flight display.

PIC. See pilot in command.

Pilotage. Navigation by visual reference to landmarks.

Pilot error. An accident in which an action or decision made by the pilot was the cause or a contributing factor that led to the accident.

Pilot in command (PIC). The pilot responsible for the operation and safety of an aircraft.

Pilot report (PIREP). Report of meteorological phenomena encountered by aircraft.

Pilot's Operating Handbook/Airplane Flight Manual (POH/AFM). Published by the airframe manufacturer, FAA-approved documents that list the operating conditions for a particular model of aircraft.

PIREP. See pilot report.

POH/AFM. See Pilot's Operating Handbook/Airplane Flight Manual.

Poor judgment chain. A series of mistakes that may lead to an accident or incident. Two basic principles generally associated with the creation of a poor judgment chain are: (1) one bad decision often leads to another; and (2) as a string of bad decisions grows, it reduces the number of subsequent alternatives for continued safe flight. ADM is intended to break the poor judgment chain before it can cause an accident or incident.

Primary flight display (PFD). A display that provides increased situational awareness to the pilot by replacing the traditional six instruments used for instrument flight with an easy-to-scan display that provides the horizon, airspeed, altitude, vertical speed, trend, trim, and rate of turn among other key relevant indications.

Residual risk. Risk left over after system safety efforts have been fully employed. It is not necessarily the same as acceptable risk. Residual risk is the sum of acceptable risk and unidentified risk. This is the total risk passed on to the user.

Risk. The future impact of a hazard that is not eliminated or controlled.

Risk assessment. An approach to managing uncertainty. Risk assessment is a quantitative value assigned to a task, action, or event.

Risk elements. There are four fundamental risk elements in aviation: the pilot, the aircraft, the environment, and the external pressures that comprise any given aviation situation.

Risk management. The part of the decision-making process which relies on situational awareness, problem recognition, and good judgment to reduce risks associated with each flight.

Single-pilot resource management (SRM). The ability for a pilot to manage all resources effectively to ensure the outcome of the flight is successful.

Situational awareness. Pilot knowledge of where the aircraft is in regard to location, air traffic control, weather, regulations, aircraft status, and other factors that may affect flight.

Spatial disorientation. The state of confusion due to misleading information being sent to the brain from various sensory organs, resulting in a lack of awareness of the aircraft position in relation to a specific reference point.

SRM. See single-pilot resource management.

Stall. A rapid decrease in lift caused by the separation of airflow from the wing's surface, brought on by exceeding the critical angle of attack. A stall can occur at any pitch attitude or airspeed.

Stress. The body's response to demands placed upon it.

Stress management. The personal analysis of the kinds of stress experienced while flying, the application of appropriate stress assessment tools, and other coping mechanisms.

Title 14 of the Code of Federal Regulations (14 CFR). Includes what was formerly known as the Federal Aviation Regulations governing the operation of aircraft, airways, and airmen.

Total risk. The sum of identified and unidentified risks.

Unacceptable risk. Risk that cannot be tolerated by the managing activity. It is a subset of identified risk that must be eliminated or controlled

Unidentified risk. Risk not yet identified. Some unidentified risks are subsequently identified when a mishap occurs. Some risk is never known.

Very-high frequency (VHF). A band of radio frequencies falling between 30 and 300 MHz.

Very-high frequency omnidirectional range (VOR). Electronic navigation equipment in which the flight deck instrument identifies the radial or line from the VOR station, measured in degrees clockwise from magnetic north, along which the aircraft is located.

VFR. See visual flight rules.

Visual approach slope indicator (VASI). A visual aid of lights arranged to provide descent guidance information during the approach to the runway. A pilot on the correct glideslope will see red lights over white lights.

Visual flight rules (VFR). Flight rules adopted by the FAA governing aircraft flight using visual references. VFR operations specify the amount of ceiling and the visibility the pilot must have in order to operate according to these rules. When the weather conditions are such that the pilot can not operate according to VFR, he or she must use instrument flight rules (IFR).

Visual meteorological conditions (VMC). Meteorological conditions expressed in terms of visibility, distance from cloud, and ceiling meeting or exceeding the minimums specified for VFR.

VMC. See visual meteorological conditions.

Index

Symbols

3P model 5-10
5P check 6-8

A

accident prone pilot.....................................2-2
Advisory circular (AC) 3-8
Aeronautical decision-making (ADM) 5-1, 6-1
Aircraft...3-4
Airport...3-6
Airspace 3-6
Air traffic control (ATC)................................6-2
Analytical decision-making 5-3
Automated flight service stations (AFSS).......6-2
Automatic decision-making............................5-3
Automatic Terminal Information Service (ATIS) 6-6
Automation 7-1
    Management...7-9
Autopilot Systems.......................................7-8

B

Baune, Helen B. 2-2

C

Checklists...6-7
Cockpit Automation Study 7-3
Controlled flight into terrain (CFIT)............... 2-4, 6-1, 8-2
Course deviation indicator (CDI)....................7-4
Crew resource management (CRM) 2-4, 6-1

D

Decision-making... 1-6, 5-3, 6-10
Decker, R. 7-3

E

Electronic flight display (EFD)...................... 7-1, 7-8
Electronic flight instrument system (EFIS)....................7-3
Environment.. 3-1, 3-5
External pressures 3-1, 3-8
External resources.......................................6-8

F

Familiarity..7-8
Federal Aviation Administration (FAA)..2-2, 2-3, 5-2, 8-1
Flight management skills 7-9
Fuller, Elizabeth Mechem..............................2-2

G

Global Navigation Satellite System (GNSS) 7-6
Global positioning system (GPS)....................7-1

H

hazard................................. 1-1, 1-2, 3-4, 4-1, 5-1, 6-1, 6-2
helicopter emergency medical services (HEMS)... 3-9, 4-2
Helmreich, Robert L. 2-4
Human
    behavior...2-2
    error 2-5
    factors related 2-2
    performance...2-1
Hypoxia...6-10

I

IMSAFE checklist.. 3-3, 6-4
Information management...............................7-9
Instrument flight rules (IFR)..........................3-5
Instrument landing system (ILS) 3-6
Internal resources.......................................6-6

M

Maximum elevation figures (MEF) 3-5
Mitigating risk..4-4
Multifunction flight display (MFD)................7-1

N

National Airspace System (NAS) 6-8
National Transportation Safety Board (NTSB) 4-2, 6-4
Naturalistic decision-making 5-3
Nighttime 3-6
Notices to Airmen (NOTAM).......................... 3-6, 6-8

O

Onboard systems..7-8

P

Passengers...6-12
PAVE checklist.. 3-1, 3-2
Perceive...5-5
Perform 5-5
Personality traits..2-3
Personal minimums...8-3
Pilot..6-12
   in command (PIC).............................3-1, 3-3, 6-5, 7-10
   error 2-2
pilot's operating handbook (POH)..............................6-7
Plan 6-11
Plane...6-11
Practical Test Standards (PTS) 7-4, 8-1
Precision approach path indicator (PAPI).....................3-6
Primary flight display (PFD)....................................7-1
Process 5-5
Programming..6-13

R

Resources 6-6
Risk 1-1, 1-2, 3-1, 4-1, 5-2, 6-1
   assessment 1-2, 4-2, 6-4
   factors 1-3, 3-7
   management........................... 1-1, 6-1, 7-10, 8-1
   matrix 4-2
Rote workmanship 7-8

S

Single-pilot resource management (SRM).............. 6-1, 8-1
Situational awareness...................................... 6-1, 7-6, 8-1
Standard operating procedures (SOPs) 3-9
Stress..3-3
   management..3-4
System safety flight training 8-2

T

Temporary flight restrictions (TFRs)..................... 3-6, 7-1
Terrain...3-5
Traits 2-4
Transcribed Weather En Route Broadcast (TWEB).......6-8

V

Veillette, Dr. Patrick R... 2-3, 7-3
Very high frequency (VHF) Direction
Finder (VHF/DF)..6-8
Visual approach slip indicator (VASI) 3-6
Visual flight rules (VFR) 3-5, 4-2, 7-3
Visual illusions...3-7

W

Weather..3-5
   minimums..8-3
Wilhelm, John A...2-4